The Daily Express Book of Herbs

The Daily Express Book of

HERBS

Harold and
Florence Priestley

Illustrated by Jane Reid

ARTHUR BARKER LIMITED LONDON
A subsidiary of Weidenfeld (Publishers) Limited

Published in Great Britain by
Arthur Barker Limited
91 Clapham High Street
London SW4 7TA

ISBN 0 213 16693 3

Printed in Great Britain by
Butler & Tanner Ltd, Frome and London

Contents

Illustrations

Foreword

This book is a simple guide to the best known and most accessible herbs; what they look like, where they come from, how they have been used in the past, and how they are used today in cooking, about the home and in medicine.

The book is divided into three parts. The first deals with garden herbs and includes many that originally grew wild but have been introduced into gardens and which may as a result of this have been modified in some way. Simple instructions for their cultivation, care, harvesting and preservation are included.

The second part lists the principal herbs found in their wild state in Great Britain, but it must be remembered that in some parts of the country these can also be grown in the garden. Much depends on the soil, the climate, the preferences of the grower and not least on the amount of trouble he is prepared to take.

The last part, though short, is important for it shows the wide variety of useful products which are derived from foreign-grown herbs, from tea and coffee to henna and liquorice.

The number of herbs dealt with is limited to those which offer the greatest rewards to the cook and the gardener, as it is primarily for them that the book has been written. Nothing is more fascinating than to see the growth of the plants one has reared, or to enjoy the new variety of flavours offered by recipes in which one's own herbs are used. It is hoped that the book will do something to encourage both gardener and cook to try out at least a few of the suggestions made in it.

In conclusion, we would like to thank Mr Albert Norman, MPS, for reading the manuscript, and Mrs D. Knight.

Introduction

A Short History of Herbs

The dictionary gives us two meanings of the word herb. To the botanist it is a plant whose stem dies down to the ground at the end of every growing season, and this includes most of the plants in our gardens. It has, however, a much wider meaning – any plant, bush or even tree of which part may be used for flavouring or garnishing food, for making medicines, drinks, perfumes and aids to the complexion. It is in this latter sense that we shall use it.

To ancient man, plants were objects of awe. Because of the unique quality of their life, so different from his own, he believed that they were part of the supernatural world, that their make-up and properties were elements of which the gods themselves were composed; that the juices of healing plants were essences of the benign gods who loved man, and those of poisonous ones were attributable to evil spirits. Hence the belief that the craft of the herbalist who held the secrets both of health and disease was founded by the gods, and that he was their deputy on earth who, with his closely-guarded knowledge, could gain their aid in the service of man.

Ancient legend is permeated with these beliefs. Osiris, the Egyptian god of plants and trees, taught man how to make wine. Isis, his wife, restored Osiris to life with herbs after he had been killed and his body dismembered by his wicked brother Set. Thoth the god-physician cured the eye of Horus the sun-god after Set had injured it. In Mesopotamia the Sumerian king

Gilgamesh, in his search for immortality, dived deep into the sea to pluck from its depths the herb which would confer it.

The power of herbs also found its way into Greek legend. Ganymede, the cupbearer of Zeus, the father of the gods, was made immortal by being given an infusion of tansy, and the centaur Chiron, half horse, half man, was believed to have cured himself of a poisoned wound by means of a herb which still bears the English name centaury.

In those early days men began to list the beneficial and harmful properties of various herbs, and so the first catalogues or herbals came to be made, in Egypt on papyrus and in Mesopotamia on tablets of clay. Thirty-six of those which were in the library at Nineveh listed some 250 vegetable remedies as well as others of mineral and animal origin.

Greek doctors derived much of their knowledge of medicine from Egypt, and founded their own schools of medicine. The two great philosophers, Thales of Miletus (639–544 BC) and Pythagoras of Samos (580–498 BC), were both pupils of Egyptian priests in their youth. The wandering physicians who went from town to town in ancient Greece practising their craft derived most of their remedies from herbs. Hippocrates of Cos (460–377 BC), the first Greek to divorce medicine from magic, used between three and four hundred drugs, mainly of herbal origin.

The greatest of all the compilers of herbals was the Roman Pedanius Dioscorides of Cilicia who lived about the time of Nero (*c.* AD 30–70). He travelled all over the known world with the Roman armies and in his work *De Materia Medica* listed more than six hundred plants as well as many mineral and animal remedies. Pliny the Elder (AD 23–79) devoted eight of his thirty-seven volumes of natural history to descriptions of medicinal plants. Claudius Galen, who lived about the same time, was a prolific writer on medicine and his work *De Simplicibus* includes a herbal (Books VI and VII).

These men set the pattern of medical knowledge for at least thirteen hundred years.

The Middle Ages were not a time of experiment. The contemporary attitude was not, as it is today, to investigate, but to accept ancient writings on medicine without question. Hippocrates had taught that there were four humours in the body – blood, phlegm, black bile and yellow bile and that any disturbance of the balance of these four would bring about illness. To this Galen had added his own theory, that people were of four types, sanguine or cheerful, phlegmatic or sluggish, melancholic or gloomy and choleric or hot-tempered. Herbs as well as people were considered to have temperaments, and to cure a person who was sluggish, for example, it was necessary to prescribe a herb which would induce in him alertness and cheerfulness. This was further complicated by the belief that each part of the body was governed by a sign of the zodiac and each herb by a planet. Thus the prescribing of remedies was a very complex business and the craft of the physician a mystery in the true sense of the word.

Alongside all these theories another doctrine developed which would now be regarded as highly superstitious but which people in the Middle Ages took seriously: that any herb which showed a similarity to the symptoms of a disease could be used to cure it. Thus the juice of the greater celandine was used to treat stomach ailments because it was the colour of yellow bile, burdock for stones in the kidneys because when its brownish-grey fruits were dried they looked like small stones, the walnut for headache and insanity because its kernel resembled the convolutions of the human brain, and St John's wort for healing wounds because when the flowers were crushed they produced a dark red liquid which resembled blood. This theory was called the Doctrine of Signatures and more examples of it will be given in the sections dealing with individual herbs.

Superstition is hard to dispel, and as far as herbs are concerned it has survived in many places to this day. It flourished alongside orthodox medicine for centuries and was intimately concerned with the struggle between good and evil as well as with relations between human beings and the spirit world. Herbs were part of the stock-in-trade of witches who, besides using them to cure diseases and to perform their magic, were said to anoint themselves with aconite and belladonna to enable them to fly at night. To ward off their spells a whole list of herbs with protective virtues, including mallow, angelica, rue and chives, was drawn up.

Among all these superstitions was mingled a large amount of genuine knowledge, much of which was owed to the Arabs who had read the works of the Ancients in translation. Their expertise was brought to Spain as a result of the Moorish invasions of the eighth century, and from there knowledge spread throughout Europe and eventually to England where physic gardens were cultivated in monastery grounds and by the owners of castles and large houses. Knowledge of the herb as a curative agent persisted in the home of the humble peasant. Experience taught him which herbs had effects on which parts of the human body and he gave them the familiar names by which so many are known today – wormwood to remove tapeworms and eyebright for curing eye infections are popular examples.

Herbals came to be produced in various countries and a knowledge of herbal remedies, as well as science, astronomy and other branches of learning, was encouraged by the invention of printing. Between 1551 and 1568 William Turner produced his *New Herball*, and in 1597 John Gerard published his *Herball or Generall Historie of Plants*. It was followed by John Parkinson's *Theatrum Botanicum* in 1652, the book which did more than any other to make the common people better acquainted with the uses of herbs. There is hardly a book published on the subject today which does not quote from one of these. They all included a certain amount of astrology, superstition and folklore, and many old doctrines such as those of the humours and of signatures were not entirely abandoned. Unfortunately, in those days there was no such thing as a controlled experiment and some recommended 'cures' probably did harm through overdosing but, when disease was rampant and death common, as it was in Stuart times, both were considered to be God's will.

Although English peasants were able to grow many kinds of herbs and to cull others from field and hedgerow, the wealthier classes were also able to buy from traders who had ventured along the routes through the Mediterranean and the Middle East where more exotic products such as ginger, pepper, cloves, cinnamon and nutmeg were available. Later, with the discovery of America in the fifteenth century, new lands were opened up to them, resulting in the importation of tobacco and cocoa and the discovery of many new plants found to have healing powers. When the New World was colonized, more and more of these formerly unknown herbs came to be used and arrowroot, slippery elm, ipecacuanha, cinchona (quinine), quassia, sarsaparilla and witch-hazel, all originating in the Americas, are now household names, while some American species such as the sunflower, the passion flower and lemon verbena are grown today in English gardens.

Both in the East and in America certain herbs were used because of their powers to affect the mind and the nervous system. The American Indians chewed the coca leaf to give them a feeling of security and contentment, and from this plant cocaine is extracted. The Indian hemp (*Apocynum cannabinum*) is still chewed and smoked in various forms in the Near and Far East, and illicit trading in its derivatives is carried on throughout the world together with a vast commerce in heroin which comes from the white poppy (*Papaver somniferum*) grown in many parts of Asia from Turkey to China. The fermentation of vegetable juices produced in Europe and elsewhere has given us beer, wines and spirits in great variety. Tea (from China and other eastern countries), cocoa and chocolate (from tropical America and Asia) and coffee are all herbal imports from abroad. Tobacco, brought into Britain during the reign of Elizabeth I and of which a small quantity is grown here today, was once believed to have many health-giving virtues. An Elizabethan rhyme runs:

My head and brains, joints and veins,
Back and reins from all pains
It doth well purge and make clean.

This optimistic comment seems sadly ironic to a modern reader who needs no reminder of the long-term harmful consequences of the habit.

The Place of Herbs Today

In spite of the many varieties of herbs which are native to the British countryside and the wealth of traditional knowledge about their use in cooking and for medicinal purposes, it is only in recent years, mainly since the last war, that interest in their cultivation and use in the home has revived in Britain to any extent. Every medieval peasant had a small herb-patch in his plot of land and yet nowadays among the profusion of gardening catalogues there are few publications in which herbs get much attention. However, there are signs of change; in particular the virtue of herbs in cooking was, until recently, largely disregarded by the ordinary British cook, but we are of late finding that almost every large store stocks rosemary, bay, balm, caraway, basil, sage, dill, garlic, marjoram, tarragon and other well-known culinary herbs. British taste, however, has to undergo a good deal of re-education before everyone emerges from the meat and two veg stage and is able to enjoy the finer varieties of taste that herbs can give. The growth of foreign travel and the increasing number of tinned and frozen foreign foods widely available in our supermarkets are doing something in this direction.

Today, as in the past, it is normally enough for the herb-grower or user to be familiar with the general qualities of each herb for culinary and medicinal purposes: that one is particularly good for use in salads, while another is better for stuffing fowl or in soups; that one is a remedy for stomach ailments while another benefits the kidneys and so on. However, modern science has gone much further in analysing, isolating and extracting the vital active principles in herbs in order to make more effective use of their natural curative properties. Though it is obviously beyond the power of the small herb-grower to isolate these substances an appreciation of their importance does not come amiss.

The most important and most talked of herbal extracts are known as alkaloids. All these contain nitrogen and in quantity are more or less poisonous to man. The first to be discovered (1803) was morphine which is derived from opium which in turn comes from the white poppy (*Papaver somniferum*). After morphine came strychnine (1819) from the *Nux vomica*, the nut borne by an East Indian tree; quinine (1820) from the bark of the South American cinchona tree; coniine (1827) from hemlock; and nicotine (1828) from the tobacco plant. Some of these and many more recently discovered are, though harmful in excessive doses, used widely in medicine.

A second derivative is the herb's essential oil, a substance which has more or less power to kill germs and is an antiseptic. Among these are anethol from anise, apiol from parsley, thymol from thyme, and the oils of lavender, linseed, peppermint, rosemary, garlic, as well as many others which are as yet unnamed. These often have a more profound effect as medicines than any other part of the herbs and some are used in the home for flavouring.

Besides alkaloids and essential oils some herbs contain substances named glycosides, poisonous to man but which in minute doses are used medically and as laxatives. One of these is digitalis, which is derived from the foxglove. Other active principles are tannins which help to promote healing, mucilages which, because they form a jelly-like substance with water, are used for bronchial troubles; acids, sugars, and last but not least, vitamins.

The methods of the ancient herbalists have by no means been entirely forgotten, and survive alongside the more orthodox approach of the modern chemist. One branch of medicine known as homoeopathy still uses the common garden herbs as drugs. The homoeopath administers minute doses – often in solutions as weak as a million to one – to cure illnesses whose symptoms the herb would cause if given in larger quantities. Thus the most poisonous substances such as aconite (monkshood), belladonna (deadly nightshade) and digitalis (foxglove) are rendered curative. In this department of medicine the wild chamomile is used for nervous and digestive troubles, the autumn crocus for gout and bad circulation, bittersweet or woody nightshade for colds and other complaints caused by exposure to damp, and rue as a lotion for bruises.

The Herb Garden

From the very earliest days when man realized that he could uproot a plant and set it down to grow in another place, wild herbs have been introduced into gardens, and many adapt easily to the transition. Some species are, however, too small and frail to be cultivated successfully and others require the competition of their neighbours in the fields to prevent them taking more than their due space. There is nevertheless an extensive range of herbs which are suitable for domestic cultivation, and since most gardeners will want to devote only a small patch of their garden to herb-growing, they will have to make a choice as to which particular kind of herbs they want to cultivate. Some may want to go in for decorative herbs, others may favour culinary or salad varieties, while others may prefer the herbs which are noted for their scent and attractiveness to bees.

There is no limit to the variety of herbs which come into bloom at different times through the season. Most flowering herbs, especially the smaller varieties, tend towards the purple and the blue, but others, such as the pinky-red bergamot, the tansy with its yellow buttons and the white spikes of the winter savory bring diversity to this. They vary in size from the creeping thyme, the periwinkle and pennyroyal to the tall mullein with its velvety greyish leaves, its yellow flowers climbing up the stalk, and the elecampane with its shaggy yellow blooms. The foxglove has a beauty which belies its poisonous nature. For variety of foliage few garden plants can rival herbs, ranging as they do from bluish-green to grey-white to form contrasts as vivid as any in other parts of the garden.

The following is a basic guide to the kind of arrangement that will show each plant to its best advantage. A border of chives or one of the thymes can be backed by medium-sized flowering or leafy herbs such as cotton lavender, hyssop, dwarf or Old

English lavender, marjoram, savory, southernwood, one or more of the sage family, sweet cicely, tansy or valerian. The taller plants should be placed behind all these, for example angelica, lovage with its high yellow umbels, elecampane, sunflower, fennel or mullein. A fully-grown rosemary bush or even a bay tree would make lovely permanent fixtures here.

The cultivation of culinary and salad herbs is also extremely satisfying. Only those who have been able to go out into the garden during the growing season and choose flavourings for roasts, stews and especially salads straight from the plant know the pleasure this gives. For salads lettuce is usually the base, but there are others such as chicory, corn salad, endive, dandelion and Good King Henry (a vastly underrated plant formerly cultivated as a pot-herb, still valued in some parts of England and known as 'poor man's spinach') whose young leaves are also suitable for salads. Savory, marjoram, alecost, tarragon, sage, sweet cicely, rosemary and tansy all mingle nicely with other flavours if used in small quantities. The final decision depends not on what the book about herbs says or what is in fashion, but on availability and individual taste; there is infinite scope for change and ingenuity.

Those who prefer the herb garden for the scent it gives should plant hyssop, the lavenders, chamomile, the marjorams, thyme, savory, pennyroyal, southernwood and rosemary. Bees tend to linger more on scented herbs and the sight – and sound – of hundreds of bees hovering over a lavender hedge in the flowering season is something to be remembered. Marigolds, mallow and sage also attract them.

Few garden flowers can rival the bergamot for scent, and this is only one of the many herbs which, due to its perfume, attracts bees and when dried may be used for making potpourri. Lavender, rosemary and basil have their own characteristic scents, lovage reminds one of strong celery and the sweet cicely plant of aniseed. Southernwood, or ladslove as it used to be called, is a favourite in the country garden and the Elizabethans were familiar with the fragrance of a chamomile lawn when it was walked over or cut. One of our own most poignant memories is of a small herb garden during the last war when a flying-bomb dropped, destroying the house to which it belonged. During the hours which followed the explosion the beds were crushed flat by the boots of the rescue squads, but in the midst of all the dust and devastation, the scent which rose from the bruised and broken herbs is, even after all these years, unforgettable.

Methods of propagating and growing vary with the character of the plant. Many perennial herbs which have been brought into cultivation from the wild state are naturally hardy and some, like the mints and tansy, throw out widespread underground shoots and have to be prevented, by trimming root and branch, from taking over more than their share of space. Most herbs will flourish in normal soil, neither too light nor too heavy, too clayey or too sandy, but it is possible with the addition of composts, peat, lime, straw, humus and other agents to cater for the special requirements of some herbs. In this respect there is nothing about herb-growing that is different from tending any other kind of garden.

In addition to spreading through root growth, some herbs may be propagated by earthing up so that roots form on side shoots, then splitting off the new growths. Cuttings are taken in the usual way, by taking off side shoots with the 'heels' if possible, putting them firmly into the ground or into pots down to about half their length, which should normally be not more than about five inches.

Annual herbs are a little more difficult. Some, like borage and mullein, will multiply by throwing or dropping seeds at random over the beds and as these grow up in spring they can be carefully transplanted. In this way there will be no shortage of new growths, and we have known both mullein and borage sow themselves in our own garden repeatedly for more than a quarter of a century. The same is the case with the Good King Henry plant, whose seeds take a considerable time to germinate, sometimes lying under the soil surface for a year or more but, once established, this so-called weed springs up in odd nooks and crannies and can be transplanted at will.

Not every flowering annual or biennial is, however, so prolific. One may, for instance, gather the seeds of angelica or another similar plant which bears umbels, by enclosing the top of the fruiting stem in a muslin bag tied at the neck to catch the seeds as they fall. But many herb seeds are apt to deteriorate if not carefully cultivated and the best supplies of most of them can be obtained only from reputable herb seedsmen.

Care should be taken in sowing. Most seeds may be sown in the open when the temperature is right, in April or even as late as mid-May, but the soil must be friable and not likely to cake in hard lumps on drying out. But to sow even a small number of varieties a seed bed is needed, and in the small garden there is often no room for many beds. The only alternative, and a good one, is to raise most seedlings in pots or trays indoors, using a good compost, and to transfer them to their final growing-place afterwards. The percentage of successes in some cases may be smaller, but in a small garden only a few plants of any one variety are usually required.

Another snag for the small gardener is the amount of room needed in the herb bed. Perennials are usually greedy for space and should therefore be given rather too much room than too little. Even then it is important to keep the plants each within its own territory, hoeing frequently between them to keep down weeds and applying dressings of fertilizer in spring and after the first cutting and manure at less frequent intervals, about once a year.

Annuals may be grown between perennial herbs if there is room, but the demands of the latter for space and the need to hoe between them makes the mingling of the two more difficult. It is not a bad idea for the beginner to treat his annuals as an experiment, at least for the first year, and to put those that he has grown from seed either in a small bed by themselves or in vacant spaces in the open garden. One should not be deterred by the fact that they are more difficult to grow, for many of them have qualities not found among hardier plants. Nothing can act as a substitute for the sweet scent of basil, and sweet cicely comes a bad second to the more delicate anise.

Harvesting and Preservation

During the growing season herbs can be enjoyed fresh, and some evergreens such as rosemary, sage and thyme may be collected even in winter, but it is always a good idea to have some dried in the kitchen. If herbs are to be preserved it is important to retain as much of their natural fragrance and flavour as possible.

Before preservation, however, the plants have to be harvested, and for this there is in every case what we call the optimum point of growth. For leaves, this is just before the plant flowers; for flowers, just as they become fully open; and for roots, when the tops of the plant are beginning to die down. Some, such as tarragon, may be good for two pickings of leaf; others grow so profusely that their leaves may be gathered in small quantities at all times throughout the season.

There are two ways of preservation, drying and, a more recent practice, freezing. For drying there are ten golden rules that should always be kept in mind. The first six of these also apply to freezing.

1 Herbs should be gathered on a dry day when there is no dew on the plants.
2 They should be handled gently and as little as possible so as not to bruise the leaves.
3 Only as many should be picked as can be dried at any one time. Herbs left to go limp and dried later will be no good.
4 No part of a plant should be chosen if it is damaged, dirty or in any way imperfect.
5 Not too much should be taken from any one plant. No more than one-third should be taken off perennial plants. Annuals are different since they will not come up in the next season. They may be cut down to the ground or taken out once harvesting is done.
6 Only the part of the plant that is needed – flower, leaf or branch – should be taken.
7 Herbs that are being dried should be kept out of the sun because sunlight destroys their flavour.

8 For drying they should be put in a really dry place. Any moisture whatsoever impedes drying and encourages the formation of mould.
9 The air in which they are drying should be kept circulating and they should be turned frequently, at least once every twenty-four hours.
10 Different kinds of herbs should not be dried at the same time in the same confined space.

Having kept these points in mind, drying may be started. It should be noted that leaves in drying are going to lose 75–80 percent of the water in them and roots a little less. Therefore a temperature warm enough to dry but not so hot as to cook them is needed. The temperature of a warm summer day is best – say 75–80°F (24–26°C) and a little higher than this for the first twenty-four hours if it can be managed. A warm room is ideal, an airing cupboard comes second – but for the linen's sake evil-smelling herbs such as valerian should never be put there.

The best drying apparatus for the home grower is a tray with a bottom of muslin, cheesecloth or hessian which can be put on supports allowing circulation of air underneath. Next best is a tray with sheets of newspaper laid on it or, if in an airing cupboard, shallow cardboard boxes lined with paper which will stand crosswise one on another. The disadvantage of the last two methods is that frequent turning is essential.

Some herbs, notably thyme, tarragon, rosemary and mint, are best dried on the branches, and these should be laid side by side on the trays and turned frequently. Sometimes they are hung on the branch and dried that way. In the case of other varieties, the leaves or flowers are stripped and spread. When dry they will be brittle enough to snap, and can then be rubbed and stored, preferably in airtight glass containers in a cool, dark place.

For freezing there are two methods. The first is to wash the herb and then, having shaken off the drops, take off the leaves and put them into the freezer on a tray as is done with certain soft fruits. Leave till firm, then take out, put into bags and return. When taken out of the freezer for use, small pieces may be broken off and used in salads if needed. The second method is to chop the herbs finely, mix them with water and freeze in blocks. In this way they will be suitable for use in soups, gravies and sauces.

The preservation of seeds is simple, as they usually dry on the stalk. They can be hung over newspaper or the heads may be enclosed in bags to catch the seeds. But they, like the leaves, should be kept in airtight containers or they will deteriorate.

Herbs in the Home

Herbs have a great number of different uses in the home, from simple domestic remedies to cooking: these are dealt with in the main section of the book. They also have a number of less familiar uses, such as in wine-making and toiletries.

Angelica, balm, coltsfoot, cowslip, dandelion, elderflower, elderberry, marigold, parsley and sage may all be used for wine-making by fermenting with yeast and adding other ingredients. Any good wine-making book will provide recipes. Other herbs such as rosemary may be used to flavour alcoholic beverages: a pleasant drink can be made by bruising and steeping it in white wine for a few days, then sieving and re-bottling. Borage is one of the best for floating on punches and non-alcoholic drinks; others are mint, balm and verbena. The most popular home-made herb beer is made from the young shoots of the stinging nettle. But for those who think of making home-made wine from their own produce, it should be remembered that herbs are light in weight and that 4 lb of sage, 1 lb of parsley or 3 quarts of marigold flowers are needed for every gallon, which may be more than their small plot can produce.

The simplest beverage that can be made from herbs is the tisane, or herb tea. There is no doubt that herb teas were drunk regularly long before tea as we know it was imported from the East, for it is inconceivable that our forefathers did not appreciate the value of a hot drink, especially on cold days. Some of the herbal flavours are quite as satisfying as that of Indian or China tea which is, after all, an acquired taste.

The instructions for making a herbal tea are simple – just pour boiling water over the chopped fresh or dried herb and leave for five to eight minutes. A general guide is two teaspoons to a pint, but this can be varied according to taste. Sweeten if necessary with sugar or honey.

Oils are extracted on a commercial scale from many herbs and some of these can be bought at the chemist. Others are used in the preparation of medicines. The true herb oil cannot be made in the normal household, but herbs such as rosemary, basil, marjoram, thyme, lavender, savory, sage and lemon verbena may be crushed either separately or together to blend flavours, then left to steep in vegetable oil for some days, frequently shaken, and at the end of the period the oil will have absorbed their scent. It can then be sieved and bottled. Used as bath oil it may tend to be greasy and leave a film, but a rather novel experiment may be tried. Choose the herb or herbs which have your favourite scent, and treat them as suggested. When ready to use, with no more than one or two drops on the palm of the hand, rub the hands together covering both back and front, then try rubbing on the arms and shoulders. The Ancients were fond of anointing themselves with oil. The process described here could hardly be called anointing but it is worth trying in small quantities, and the scented oil used in this way lasts a long time. Oil of marigold, made from the petals of the flower, is good for the feet after a bath.

A wide variety of herbal scents can be used in the bath. Either a single herb or a mixture of several can be used: some people would add clary, sage and lovage to the usual varieties. A strong infusion can be made by pouring boiling water over the herbs, allowing it to stand, then straining the liquid when required into the bath. Thus the greasiness of oil, which leaves a deposit on the porcelain or enamel, is avoided. Alternatively, and this is more economical though perhaps not as effective, they may be put into a muslin bag or into a cloth gathered, tied and then dropped into the bath or left to hang from the hot tap where the water passing over and through them will become impregnated. The same bag, squeezed out, may be used more than once.

A decoction of herbs makes a good hair rinse. Again a choice may bc made among the scented ones, bearing in mind those which have a special value for the hair – chamomile, fennel, sage and rosemary. For fair hair a preponderance of chamomile is advised, for dark hair, rosemary. Half a cup of dried herbs (a little more of fresh herbs) to two pints of boiling water, allowed to cool to the right temperature, is sufficient for a rinse.

Stronger infusions – a cupful of rosemary leaves or chamomile flowers to a pint of boiling water – will make rosemary or chamomile water if allowed to stand for a week or so, shaken daily. Those two are in themselves efficient scents and skin tonics but should be used within a few weeks to forestall the formation of mould.

Simple face packs may be made by mixing Fuller's Earth (available from chemists) into a paste with a herbal infusion of some of those mentioned, even also dandelion and nettle. Spread thinly over the face with fingertips or a soft brush avoiding the eyes, which may be covered with cotton wool. Leave for ten to fifteen minutes, then wash off and rinse the face with cold water to finish. Use 1 oz of herbs to 1 pint of boiling water left to cool before mixing into a paste. Fennel, which is an antiseptic, helps to smooth wrinkles and chamomile is an astringent for ageing skins.

A compress for a skin tonic may be made by dipping a piece of soft cloth into a cold infusion of herbs and leaving on the face for ten minutes. Suitable ones are sage and yarrow.

Herbs which can be Grown in the Garden

Aconite

Aconitum napellus Perennial

Some plants, like animals, have been domesticated, and aconite is an example of these. Wild, it grows in clumps in a few parts of England and was probably brought here from the Continent before the Norman Conquest.

Around its stalks cluster deeply-divided dark-green shiny leaves, and because of the shape of its blue flowers it is commonly known as monkshood. Easy to grow, especially by means of tubers taken from a main root, it makes a handsome addition to a flower border but one should be wary for it is very poisonous. It should be handled carefully with gloves and should never on any account be grown among edible herbs. From the roots the active principle, aconitin, a rapid and deadly poison is extracted.

Its name is derived from the legendary hill of Aconitus which stood at the entrance to the infernal regions. Here Hercules fought and captured the three-headed dog Cerberus, and the plant was said to have sprung from the foam which fell from the monster's mouth.

Homoeopathic preparations made from the flowering tops are given internally in dilutions of millions to one at the onset of colds and fevers and to minimize the effects of shock. Other preparations in the form of tinctures and liniment have been used locally for neuralgia, lumbago and rheumatism.

For the amateur, aconite, once planted, is best left to grow at its own pace, and not too freely handled. Preparations from this plant must only be prescribed by a physician.

Alecost or Costmary

Chrysanthemum balsamita Perennial

This old English plant is now sadly neglected in the herb garden, because of its wayward and untidy growth, its tendency to throw out underground shoots, its pale, rather unattractive leaves, its poor little clusters of yellowish florets rising about a couple of feet, and its complete rejection in modern times as a medicinal plant.

But let the passer-by crush but one small leaf between thumb and forefinger! No wonder the Elizabethans strewed it on the floors of their rooms, that Linnaeus gave it the name *balsamita* (fragrant plant) because of its unique scent, that the French named it *Herbe Sainte Marie* (herb of the Virgin) and that it was known in English as costmary (from the Greek word *costos* – an aromatic plant).

If rather wayward, it is easy to grow, can be cultivated in almost any soil and is tolerant of most climates, preferring dryness and sunshine. But it should be controlled by splitting off side shoots where necessary and surplus growth can be divided from the main plant and thrown away when it has become too ugly or too woody.

All we can say is, try it – dried, as an ingredient of potpourri or, fresh from the garden, a small amount cut finely and added to a salad. The home brewer might try imitating his forbears, bruising a few leaves and allowing them to infuse in an odd pint or two of his latest brew. He may like it – or he may not!

Angelica

Archangelica officinalis Biennial

A lush green plant with large indented leaves rising from a single rosette on long hollow stalks, angelica can be one of the most prominent features of the herb garden. It prefers shade, good damp soil, and throws up thick stalks ending in large umbels which rise from opened sheaths, bear clusters of small light-green florets and seed abundantly.

A biennial plant, best grown from seed, it creates its ground rosette in the first year, flowers and seeds normally in the second, then dies. It can be kept a third year and even longer if not allowed to expend its energy in flowering. But if it does not flower it bears no seed; therefore, for the small gardener, it is as well to have two plants, the first seeding while the second is growing ready to bloom and seed the following year.

Seeds may be sown in open ground in August or September where they may survive the winter and come on early the following spring, though this is by no means certain in every case. It may be better to sow in pots or trays in early spring and transplant the seedlings out when very young into their growing-space, preferably at the back of the herb patch, for the plant grows rapidly, needs about three feet to spread itself and may, if allowed, throw up umbels to a height of six feet.

Its name indicates the esteem in which it was held by the Ancients; it became linked with the spring festivals of the Annunciation and the day of St Michael the Archangel which fell at about the time when it came into flower.

All through the growing season some part of the angelica plant is usable. As early as May the juicy young shoots are springing up and these may be taken off and candied. It is important to cut them just at the right time before they become stringy, for growth is very rapid. If tender enough they can be cut up and cooked with rhubarb or added to jam to vary the flavour. In some parts of the world the stems are baked or boiled and eaten as a vegetable.

In June the fresh green leaves may be used sparingly as an

Angelica
Archangelica officinalis

ingredient in salads or dried for making herbal tea. The roots, dug up, sliced and dried, have a stronger odour than the rest of the plant. Boiled, with honey added, they are an old household remedy for flatulence and digestive troubles.

Angelica has always been one of the chief flavouring agents in confectionery and drinks and in the preparation of vermouth and chartreuse. Candied angelica may be bought for decorating confectionery. The candying can be done at home though the method is complicated and possibly uneconomical for very small quantities.

Anise

Pimpinella anisum Annual

For the English garden anise is a rather tender plant and not likely to produce its most important part – the seed – except in mild, sheltered areas, or in the south of England. Small, growing to a height of about eighteen inches, it produces tiny white flowers on branching umbels.

It is grown more widely in the region of its origin, southern Europe, where the seeds ripen and are distilled to produce the fragrant syrupy oil which is the basis of aniseed flavouring. Travellers may have sampled the popular alcoholic drinks, to which aniseed gives its flavour, the French *pastis*, the liqueur *anise*, produced in France and Italy, and the Greek *ouzo*.

It is best grown from seed sown in April in light, well-drained soil and thinned out, leaving plants about six inches to a foot apart where they may flower in late July or August. When the greyish-brown seeds have ripened – if they can be got to ripen – the plant is cut or pulled up and dried on trays. The seeds may be crushed and put into sachets which are useful, especially if the herb is mixed with dried lavender flowers, to put among the bed linen while it is being stored. Boiling water may be poured over the seeds to make a drink (2 teaspoonfuls to ½ pint) which acts as a relief to colds and bronchitis, especially for children. For those who like the flavour it can be used, as it was by the Romans, as a powder to flavour cakes or in puddings and sauces.

Balm
Melissa officinalis

Balm

Melissa officinalis Perennial

Balm will flourish even in poor soil, and may best be propagated by the splitting of roots, though it can be grown from seed. It reaches a height of more than two feet and needs about three feet of ground, as it throws out long shoots which bear serrated leaves and smallish white or light-yellow flowers at the leaf junctions. The leaves, crushed, give out a strong lemony scent which persists and is not destroyed by drying.

The plant has a strong attraction for bees, hence its Greek name *melissa*, a bee. It is grown for its leaves which may be taken off before the flowering period and dried either in the shade or on a tray in gentle heat (up to 40°C).

Balm, as its name implies, has from ancient times been valued to 'comfort the heart', to calm the nerves, decrease tension, bring down the rate of heartbeat and promote sleep. It may be drunk as tea using fresh or dried herbs (fresh, 1 oz; dried, 2 teaspoons to a pint of boiling water), and sweetened with honey, or it may be added to twice its quantity of tea to make a refreshing drink. In the potpourri jar it gives zest to the scent and helps to make it last. Finally, try pinching off a few young leaves to chop up and add to salad.

Sweet Basil
Ocimum basilicum

Basil

Ocimum basilicum – Sweet Basil
Ocimum minimum – Bush Basil Annual

More basil is now bought dried in this country than is grown, for it is native to warmer climates where it may survive the winter, and is therefore rather difficult to grow. But for the sheer pleasure of seeing it grow and for the spicy odour, it is worth a trial. Seeds of sweet basil may be sown in heat in early spring or under glass in April. The seedlings may be transplanted either into pots or into the annual herb bed about a foot apart. Sweet basil in a pot will reach a height of about two feet, growing from a single main stem, its branches bearing spurs of white flowers and green scented leaves for which the plant is raised. Considering the relatively small return, drying may not be worthwhile for the person who raises no more than one or two plants.

Its use has always been more widespread in the East where, in India, it is considered sacred to Krishna and Vishnu, and in Mediterranean countries, where it is profitable to harvest it. The bush variety, smaller and more hardy than sweet basil, with much the same scent and flavour, was widely grown in England during Tudor times, and today may with luck survive a mild winter. In days before refrigeration the plant was used as a food preservative and to mask the flavour if meat became 'high'.

With the increase in continental travel the British public is now beginning to appreciate its flavour, to use it with other herbs as an addition for soups and sauces, and to liven up a salad.

Tomato and Basil Salad

Wash and slice tomatoes thinly and place in layers in a dish, interspersing finely chopped or shredded basil leaves between the layers.

Sprinkle with lemon juice, a little oil, if liked, and salt and pepper. Leave an hour before serving with cold meat or fish.

Bay
Laurus nobilis

Bay

Laurus nobilis Evergreen tree

In warmer climates the bay tree will grow to a height of about sixty feet but in Britain it rarely exceeds twenty-five. It is best propagated by taking young shoots in August, but they are not easy to strike. However, a fully-grown bay tree will drop seeds which, if undisturbed, may occasionally take root and the seedlings can then be potted. Thus it may be possible to get a bay seedling from a kind friend or neighbour. When very young it is best to keep it in the pot so that it can be taken indoors in cold weather, for the young plant is tender and easily killed. Once it is firmly rooted and has become some fifteen to eighteen inches high it will then grow quickly, sending out fresh green shiny shoots with dark green leaves which can be taken off and used, or dried for preserving. Once established it will stand up to the most severe weather, and if it grows too large for the small garden can be trimmed, losing none of its attractiveness.

Bay belongs to the laurel family and may easily be mistaken for the poisonous cherry laurel, so it is vitally important to be sure of its identity before using it.

It was known in ancient Israel (Psalm 37), and because of its delicate scent made up the wreaths which were used to crown victors in athletic and other contests in both Greece and Rome.

The leaves were once used medically to bring on perspiration, to reduce fevers and increase appetite, but today their main use is as a flavouring for soups and meat dishes. They are, like most evergreens, easily dried on the stem, when they go a light brown and tend to curl. They can, however, be pressed flat and stored.

Bergamot
Monarda didyma

Bergamot

Monarda didyma Perennial

Bergamot is not a native of Great Britain nor is it a renowned medicinal plant, but because of its whorls of scarlet flowers, its attraction for bees in the herb garden and its sweet scent which makes it a most important ingredient of the potpourri jar, it must be included. Originally found wild in the swamp lands of North America, it was first introduced into this country in the eighteenth century.

During the War of American Independence (1775–1783) many colonists used the young leaves of the plant to make a drink rather than buy the Indian tea which they had been compelled to import by way of Great Britain. For this reason it came to be known as Oswego Tea after one of the North American forts. The tea is said to be soothing to the nerves and an aid to digestion.

Today cultivated varieties of bergamot may be found in white, pink and mauve as well as scarlet. The plant grows well in most kinds of soil to a height of between two and three feet, but flourishes best in part shade with plenty of moisture. It is propagated by dividing the roots, and this is best done every other year, using the outer shoots for replanting, as the inner part of an old plant tends to die.

Treated in this way the bergamot adds variety to the yellows, blues and purples of the herb garden, and can be used as a garnish for wine cups and punches, and as a delicious ingredient in salads (when used very sparingly).

Borage
Borago officinalis

Borage

Borago officinalis Annual

Here is a herb which, kept under restraint, will behave well, but if left to its own devices will make very free with your growing space, shooting up from seed in every nook and cranny. Seeds sown in spring grow rapidly and flower in July and August. The hollow stem grows to a height of about eighteen inches and from it branch rough, hairy, oval leaves, the lower ones on short stalks. Clusters of bright blue flowers top the main stalks and hang like small stars, each bearing a tiny black cone-shaped anther. The flowers fall off rapidly and strew the ground with tiny blue specks.

Once established it is easy to cultivate, broadcasting its brown seeds so that in spring the small deep-veined leaves open on the surface of the soil even where they are not wanted. Seedlings can be moved in their own clump of earth when very small and placed in a group with others in the herb bed.

A mild, friendly herb, it has been held in high esteem by man from time immemorial.

I, Borage
Bring always courage

runs an old rhyme. Bacon and John Evelyn sang its praises. Put into wine, says the herbalist Gerard, it makes men and women 'glad and merry and drives away all sadness, dulness and melancholy'.

Though it is not used today in medicine it is still a remedy in the home. Used as a tea it 'gives grace to the drynkynge' (Gerard) and is good for catarrh. The leaves, bruised, give out a mild odour of cucumber and for this reason it is used as a flavouring in cool drinks and can be put into salads and sandwiches.

Caraway
Carum carvi

Caraway

Carum carvi Biennial

The old-fashioned seed-cake is beginning to be popular again and caraway seeds are appearing on supermarket shelves. But why do we not grow them?

The plant grows wild all over Europe and parts of Asia, reaching a height of some eighteen inches to two feet, with fine feathery leaves and umbels of small white flowers which come out in June. In England it is best sown in August or September to flower the following year and, apart from regular hoeing, needs little care. The small curved fruits, or seeds, as they are called, may be taken from the plant on the flower-heads and dried either by hanging over paper, which will catch them, or in paper bags tied at the neck. Their spicy smell and taste have made them among the best known of all herbs.

Caraway was familiar to the Ancients and the name comes from the Arabic *Karawya*. Its seeds were a delicacy in medieval England and were eaten with apples, raw or roasted. In Shakespeare's *Henry IV*, Part 2, Justice Shallow asks Falstaff to join him in 'a pippin and a dish of caraways'.

The seed is best known today as a flavouring to be baked in or scattered over cakes, though in other European countries it is used to flavour soups, cheese and bread.

Medicinally, the seeds are used to make a tea to stimulate the digestion, the liver and gall bladder, and to increase the flow of milk in nursing mothers. It is specially useful, often combined with chamomile, for flatulent colic in young children, being one of the constituents of gripe water.

Chamomile
Anthemis nobilis

Chamomile

Anthemis nobilis Perennial

Few herb gardens exist today which do not contain this small sweet-scented plant, 'the herb of humility', which retains the fresh green colour of its foliage in the dryest of seasons and produces its small white yellow-centred daisy-like flowers.

Growing close to the ground, chamomile throws out runners in all directions from the main plant, and these readily root or can be taken and planted elsewhere. It is said that Drake, if indeed he played bowls while the Armada was approaching up the Channel, played on a chamomile lawn. Such lawns are common and can be made either with small half-rooted runners planted about six inches apart and allowed to spread, or by sowing seeds in a tray and planting out the seedlings when large enough. Today there are varieties specially grown for lawn-making. If double flowers are wanted, it is preferable to buy the plants from a herb farm. If they are watered well, and the ground is weeded regularly after planting, they will probably cover the ground completely by the end of a growing season. In the first year such a lawn should be cut carefully, taking off the upper shoots and flowers, as this assists the spreading process. In the second season when the plants are established they may be cut with a mower with the blades set as high as possible – and what a sweet scent they give off either when mown or walked over!

On the other hand, they may be required for their flowers which are picked off without their stalks when they are in full bloom. For this purpose the English or double chamomile is preferable.

The daisy-like centre of the single chamomile rises from the stalk like a small yellow ball, and for this reason the Greeks called it *kamai melon* – ground-apple – from which the modern name is derived. Its name in Spanish is *manzanilla* – the little apple. English herbalists valued it not only as a refreshing addition to bath water but as an infusion or tea. This is recommended as a relaxant and sedative, for relieving cramps and menstrual pains and for easing a sore throat. It is made in a closed vessel

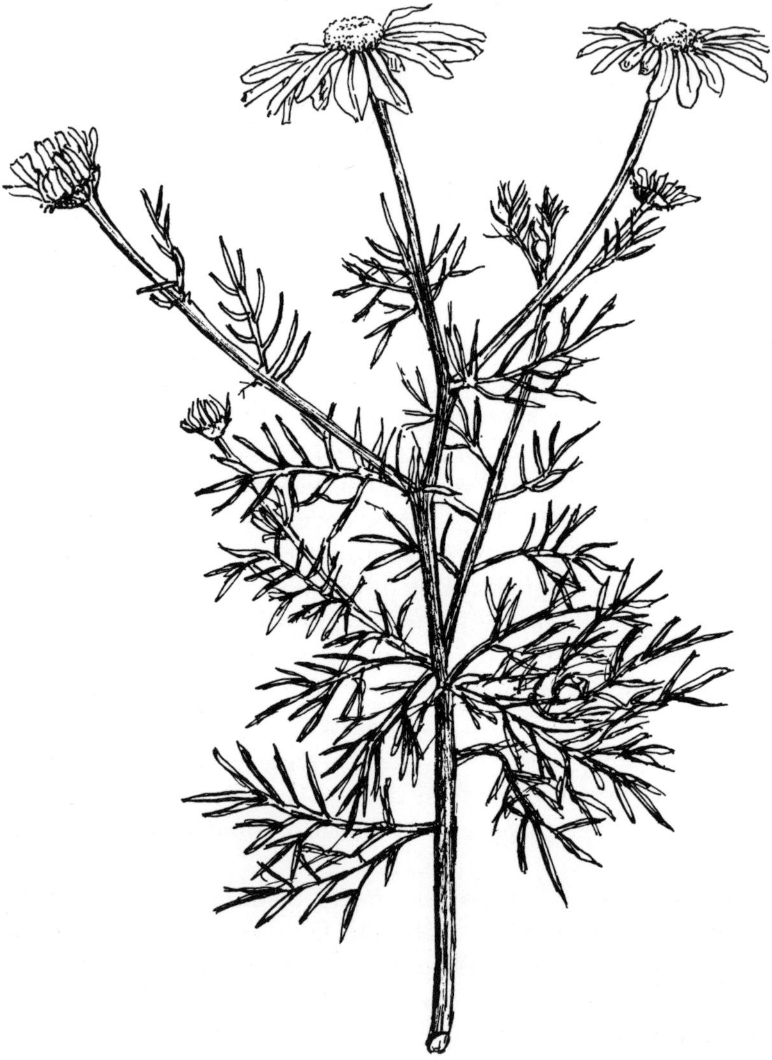

German Chamomile
Matricaria chamomilla

by allowing the flowers to stand in water fresh from the boil for about ten minutes. Externally it can be used hot as a fomentation or on wounds and burns. The flowers can be dried and preserved in the usual way.

A second variety of chamomile, the German or *matricaria chamomilla*, is a wild plant, a modest little weed growing to a height of about two feet in cornfields, on waste ground or by the roadside. It has small white flowers much like *anthemis nobilis* except that its petals bend sharply backwards. It is much less aromatic with a rather acrid apple-like scent and a bitter taste. This is the chamomile so highly prized by the homoeopathic physician for illnesses in which extreme restlessness, especially in children, is one of the main symptoms.

'When the husband complains of the wife's being cross and irritable,' said an American doctor, 'give him a dose of chamomile. It works.'

But why only the husband?

Chervil

Anthriscus cerefolium Annual

The window-box gardener would welcome this little plant which, though its medicinal value is almost negligible, can be grown from seed almost all the year round and is superior even to parsley as a culinary herb.

Growing only one to two feet in height, it is one of those tender herbs which produce light green leaves in plenty and small white flowers on umbels which should normally be allowed to flower only to produce seeds. Autumn sowing in shallow drills in an outdoor spot where sun and shade alternate will make it flourish and eventually, like borage and other seeding plants, appear in odd places year after year. An alternative sowing-time in the open is April and then at regular intervals throughout the summer. In boxes it will grow from January onwards but, once sown, it does not take kindly to being transplanted nor, if in the open, to long periods of hot sun. Nor does it dry well. It is a green herb best eaten fresh from the garden.

It will soon probably be grown more in this country. The French make much use of it, including it in their select bouquet of *fines herbes* for flavouring soups and sauces. It is the first green plant of spring and almost the last of autumn; it has a tenderer, more aromatic leaf than parsley and there can be hardly a single dish in the cook's repertory of savouries and salads that would not be improved by its inclusion.

Cod with Chervil and Parsley

- 1 teaspoon of chopped onion
- 2 oz butter
- 2 lb cod, boiled, skinned and chopped
- 1 teaspoon chopped chervil
- 1 teaspoon chopped parsley
- freshly-ground black pepper and salt
- juice of half a lemon

Fry the onion gently in the butter until tender. Add all the other ingredients and stir over high heat for 5–10 minutes.

This, like the other recipes in this book, serves 4.

Chervil and Lentil Soup

- 4 oz lentils
- 3 tablespoons chervil
- 1½ pints stock
- 3 oz butter
- 1 tablespoon cream
- salt and pepper

Soak the lentils overnight. Cook them in 1 pint of water until soft (about ¾ hour). Put through a blender or sieve. Sauté the chervil in the butter for a few minutes, then add the creamed lentil mixture. Cook for 10 minutes, then add the cream and serve hot.

Chicory
Cichorium intybus

Chicory

Cichorium intybus Perennial

It is now common to see the white, green-tipped hearts of chicory on sale in the larger greengrocers' shops and in supermarkets, but these bear little resemblance to the normal plant which grows freely in the kitchen garden. Here it springs from a rosette of spreading leaves, throwing up stalks to a height of some three feet if allowed to grow unhindered, and these bear largish flowers of a delicate blue. It is fairly common as a wild plant on light soils, especially in the south-east.

Chicory has always been valued more in continental countries than in Britain, chiefly because of its deep, thick roots which are sliced, dried in kilns, then roasted and ground for mixing with coffee. This coffee-with-chicory mixture has been imported into Britain in ground form for more than a century.

Since Roman times the plant has been used for salads and even eaten boiled as a vegetable. Growing normally, its leaves are bitter and unpalatable, though it is prized on the Continent as fodder for horses, cattle and sheep. It can be sown in open ground in May or June and thinned to about eight inches between plants later. The roots are dug up in October, trimmed, then planted to about an inch below the crowns in boxes of damp sand or soil. If they are kept moderately warm, with all light excluded, within some four or five weeks the tender white hearts enclosed in delicate light green sheaths will be ready for the salad bowl.

Chicory has no distinguished place in the modern herbal, though at one time the root, cut up and allowed to stand in boiling water, was considered a good cure for jaundice, gout and rheumatism.

Chives

Allium schoenoprasum

Chives

Allium schoenoprasum Perennial

In French, Italian and Spanish chives are called the 'little onion'. They are the ideal bordering plant with hollow grass-like leaves growing from clusters of small bulbs and producing almost solid clumps of rich purple flowers which take on a pinkish hue as they mature in July. Growing to a height of about eight inches in almost any variety of soil chives can be propagated by the division of clumps or by the splitting off of a few of the small bulblets and replanting. They will quickly multiply and form fresh clumps. All that is then needed is to keep them free from weeds and the blades of grass that sometimes tend to grow among the leaves.

The leaves alone are used, and these may be cut off at any time during the growing season. It is more common, though, to take them off as required. If allowed to grow freely the little spears tend to go yellow and wither at the ends, so it is preferable to cut them frequently, as new tender ones grow all the time. After the blooms have died the whole plant may be cut within a couple of inches of ground level and it will come up again as healthy and hardy as ever.

Salads, sandwiches, omelettes – all benefit from the addition of chopped chives, which have a milder flavour than any of the larger kind of onion. Above all, they should never be omitted from a dish of potatoes either served whole or mashed.

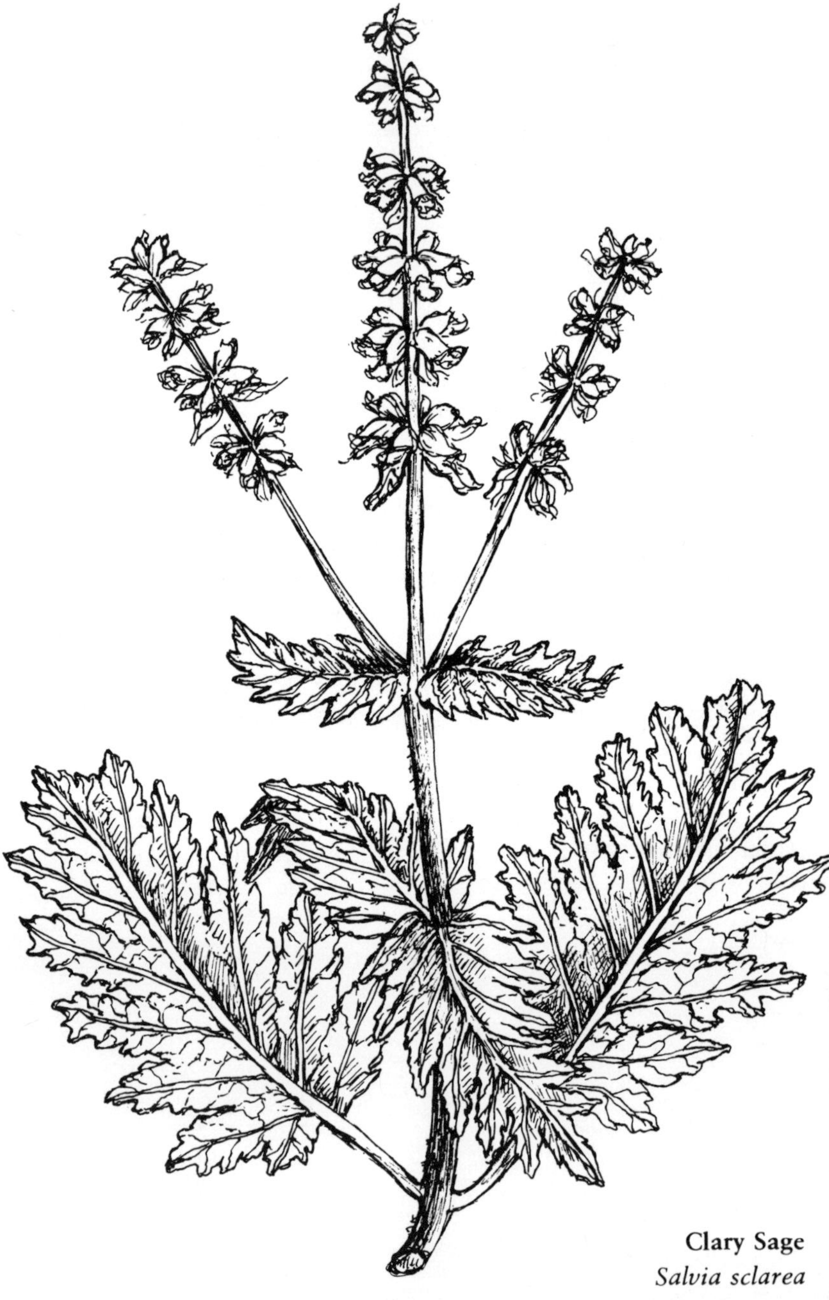

Clary Sage
Salvia sclarea

Clary Sage

Salvia sclarea Biennial

Whatever its medicinal properties, and these were once highly valued, clary should not be forgotten by the herb-grower for two reasons. It is a superb garden plant with large sage-like leaves and flower spikes rising to three feet or more and bearing small bluish-white florets. The scent is strong like that of the common sage but with a tang resembling that of grapefruit rind. It remains in the nostrils for hours, and in the memory for ever. For this reason it is best used sparingly in the potpourri jar both as a scent in its own right and as a fixative for other scents.

If grown from seed, clary should be sown in the spring and planted out in the usual way, leaving a foot all round each plant, to produce a profusion of leaves in winter and the following spring, and flower in summer. It will then die, but not without having scattered large numbers of blackish-brown seeds which can be collected if needed before they fall. When left in water these seeds produce a mucilage which was once used to take foreign bodies out of the eye, hence the English name 'clear-eye'.

The plant is not now used as a medicine though the leaves used to be boiled up with vinegar and mixed with honey to make poultices for boils and abscesses.

Comfrey
Symphytum officinale

Comfrey

Symphytum officinale Perennial

Comfrey is so well known for its healing qualities that ointment prepared from it may readily be bought at chemists and health food stores. It grows wild in England near rivers and ditches, and is specially fond of shade. It was cultivated by peasants in medieval times as a garden plant.

Growing to a height of some four feet, it needs plenty of room, and there are mixed feelings as to its beauty in a herb garden because of its large hairy leaves and its stout bristly stem. It has, however, a long flowering period, the common comfrey being in bloom from about May onwards, showing small hanging bell-shaped flowers decreasing in size as they branch, curving over from an erect or almost erect main stem. These eventually produce small nut-like fruits. The plant is best propagated by division or splitting off from main roots and establishes itself so readily that if not controlled it can become a nuisance. For the gardener, comfrey is an excellent addition to the compost heap, as it helps to promote fermentation.

As its various names, knitbone, boneset and bruisewort imply, it has great healing powers. Its roots, beaten, boiled and allowed to stand in hot water, produce mucilage, which when spread on a clean cloth may be applied directly to an inflamed area and helps to stop bleeding and to heal running sores, burns and bruises. The whole plant, mashed and laid on as a poultice, once had a great reputation for healing broken bones. The liquid left over from boiling comfrey may be used as a wash for skin complaints. A tea can be made from the roots or they may be infused in milk and water to make a cough medicine. Alternatively ½ oz of crushed root may be simmered 20 minutes in 1½ pints of milk and water.

There are other varieties of the plant including the wild tuberous comfrey (*Symphytum tuberosum*) and the large prickly comfrey (*Symphytum asperimum*) which was introduced into England in the early part of the last century as a fodder-producing plant, but without much success. Others specially cultivated for the gardener are the dwarf comfrey, a ground-hugging and weed-killing variety, a crimson-flowered variety and one with silvery leaves.

Coriander

Coriandrum sativum Annual

'The common coriander is a very striking herb; it has a round stalk full of branches, two feet long. The leaves are almost like the leaves of parsley but later on become more jagged. . . . The flowers are white and grow in round tassels like dill.' (Gerard's *Herball.*)

The herb is sown in open ground in April about half an inch deep in drills between eight and nine inches apart. It is slow to germinate, like parsley, but once established it grows rapidly, reaching a height of up to two and a half feet, and should be thinned out to between six and nine inches apart. Its seeds, which are the only part used, have a scent rather like that of an orange and fall as soon as they are ripe.

Originating in the Near East, coriander was valued both in Ancient Egypt and by the Greeks. The Romans introduced it into Britain, using it as a spice and flavouring. Today the seeds form an ingredient of many pickles, sauces and curry powders, and serve to disguise the disagreeable taste of some purgative medicines. Of itself, coriander stimulates the appetite, helps the flow of the gastric juices and is added to other medicines to relieve intestinal colic.

Corn Salad

Valerianella locusta Annual

As its name indicates, this little plant, growing to a height of twelve inches at most, is used for salads, especially during winter and spring. It is found wild on waste ground and on land which has been cultivated, especially in cornfields.

If the leaves are wanted for use in winter it is best grown from seed supplied by a nursery and sown in September, and the seedlings thinned out to about six inches apart. The shoots, which

Corn Salad
Valerianella locusta

Dill
Anethum graveolens

usually spring separately from the ground, are bright green and bear small stalkless, lance-shaped leaves which should be taken when young and crisp. If seeds are needed, a few plants may be allowed to produce the small, greenish-white flowers in late spring. One great advantage of the plant is that it can be sown almost the whole winter through, and is aptly known as lamb's lettuce. In France it has been called priest's salad or canon's salad because it is eaten in Lent.

Though its efficacy as a spring blood purifier was once valued by country folk it now has few or no medicinal uses.

Dill

Anethum graveolens Annual

Dill, though not as tall as fennel, may easily be confused with it, as the two plants are similar, and the seeds when sown should be kept well apart because of the dangers of cross-fertilization. Dill grows rapidly on a single stem, branching and bearing feathery leaves and flat umbels with clusters of small yellow flowers which soon produce numerous oval seeds. These may be collected by cutting off the flowering stalks when the seeds are ripening, and allowing them to dry off under suitable conditions. The whole plant has an odour, and the seeds a rather pungent taste resembling caraway.

It needs good, fairly rich soil and can be grown in normal conditions if protected from extensive periods of hot sun and from strong winds. It can be sown in late March or April and thinned out to about ten inches between plants.

The herb is a native of Mediterranean lands and has a long history. In the Gospel according to St Matthew Christ reproves the Pharisees for paying tithes in mint, dill and cumin and overlooking the demands of the law, justice and mercy. It was used as a drug in Greece, Rome and medieval England to stop hiccoughing, to 'expel wind and the pains proceeding therefrom', and as a protection against the spells of witches.

A century ago when a baby suffered from gripe the first remedy the mother sought was dill water. This was made either by bruising and steeping the seeds in water or by adding a few drops of the essential oil, bought from the chemist, to water. Today it is hardly ever used as a medicine. The seeds are added to vinegar when pickling cucumbers, and the leaves can be used sparingly in sauces and salads, with new or mashed potatoes and, like fennel, when cooking white fish. Chopped dill and mashed cucumber make an interesting combination when blended with butter.

Cucumber and Dill Butter

Peel and mash a small cucumber, or half a large one. Beat 1 oz butter until soft, blend with the mashed cucumber, the juice of half a lemon and a teaspoonful of chopped fresh or dried dill.

This can be potted and kept cold until needed, and makes a tasty tea-time sandwich spread between thin slices of bread and butter. It is also delicious served with cold meat.

Dill Fritters

Mix 4 oz cooked rice, well boiled and strained, with a chopped onion, 2 teaspoonfuls chopped fresh or dried dill, 2 oz grated cheese and a little frying oil. Add the yolks of 2 eggs and knead well. Form into cakes, roll in beaten egg, then in flour, and fry in hot oil, turning, until lightly brown on both sides.

Elder

Sambucus nigra Deciduous tree

The elder is a handsome tree, an ornament to the landscape with its lush green leaves and creamy blossoms, but the owner of a small garden is well advised not to try to grow one at home, for the tree grows rapidly and is land-hungry. Beneath its branches, which may grow as much as six feet in one season, nothing except the coarsest vegetation such as nettles will flourish.

In medieval times the pith was extracted from the stems which were used to make whistles, flutes and other musical instruments. Little boys used them as blowpipes and peashooters. The elder is associated more than most trees or plants with folklore and superstition. The Cross of Calvary was believed to have been made from it, and Judas to have hanged himself from one of its branches. No furniture was made from it, no fire ever kindled from its twigs, and its branches were carried aloft in funeral processions. It was the tree of protection from all evils, and often found a place by the door of the peasant's hovel.

All parts of the tree have been used; the bark as a black dye or infused in water as a purgative; the leaves for making an ointment to soothe bruises and sores and, boiled with a little linseed oil, as a poultice for piles. The flowers, collected at their best and dried quickly either in a heated pan or a cooling oven, were used for making elderflower water to tone up the skin, and elderflower tea, often drunk with peppermint and yarrow against the onset of colds, chills and influenza. The berries, simmered in a little water and thickened with sugar, served the same purpose. There are numerous recipes for wine which can be made both from the berries and from the flowers.

Indeed, the elder is so widespread and so prolific that it would seem a waste not to try some of the recipes in which it is used either as a drink, lotion or conserve.

Elderberry Jam

Boil 4 lb berries with 1 pint water, the juice of 8 lemons and 4 lb sugar. Test for setting after about 30 to 45 minutes.

When ready put into jars and close when cool.

Elderflower Wine

¾ pint elderflowers
1 gallon water
3½ lb white sugar
½ lb raisins
juice of 3 lemons
½ teaspoon yeast

Cut off the florets, press down, put into a bowl with the sugar, raisins and lemon juice; pour on boiling water and stir. Allow to cool to about 70°F (about 20°C) and add yeast. Cover and stand, preferably in a temperature of about 60–70°F, stirring daily, for 4 or 5 days. Then strain into a jar and fit trap.

When the fermenting ceases (2–3 months), allow to clear and bottle.

Elecampane
Inula helenium

Elecampane

Inula helenium — Perennial

This truly handsome garden plant is probably a native of Southern England though it is rarely found now in a wild state except in a few damp and shady places. It is more widely established on the Continent and has found a habitat in the eastern United States. It has a prominent place, however, in the English herb garden where it can be appreciated to the full.

It may throw up a main shoot as high as six feet, rising from a thick bunch of olive-green leaves, some of them as long as two feet. Smaller ones cling to the coarse stalk as it rises upwards. The large ragged yellow flowers borne at the top and on short stalks branching from the main stem make it an imposing sight as a background to the herb bed.

It may be grown from seed, but a better method of propagation is by taking part of an established plant with a portion of the root on it. Once growing it needs little care and will last for some years.

In the Middle Ages elecampane was one of the most important medicinal plants in existence.

I have a little bottle in the waistband of my britches
That is known by the name of Alycampane,
Shall make this goodly champion
Rise up and fight again.

So speaks the Doctor in the old folk play of Saint George as he comes to revive the hero during his struggle with the Turkish Knight.

The Greeks knew the plant, and it is associated with Helena, wife of Menelaus, captain of the host which fought against the Trojans. Its Latin name *inula* is believed to be a corruption of the original Greek Helen. The Romans used the root as a digestive and tonic and ate it as a vegetable, and its strange English name is a second corruption of the Latin name *Inula campana*, given to it because it grew wild in large numbers on the lowlands of Campania around the Bay of Naples.

Fennel
Foeniculum vulgare

Both in England and the Continent the root, when taken at its best from a plant about two years old, was widely used for coughs and chest troubles. The dried roots were prepared with sugar or beaten and put into beer, made into a tea or a cordial by infusion with sugar and currants in white wine, 'whereby', says Culpeper, 'it relieves coughs, shortness of breathing and wheezing in the lungs'.

Its active principle, inulin, was discovered as long ago as 1804. In Switzerland and France it is one of the substances used in the manufacture of absinthe.

Fennel

Foeniculum vulgare Perennial

One of the larger herbs, fennel in springtime throws up thick, smooth, bright—green stems to a height of five or even six feet and with them masses of feathery leaves which are an adornment to any herb bed. Its yellow florets crowd together on large flat umbels each consisting of up to twenty small clusters.

Originally a wild plant believed to be native to Britain, it still grows wild in the south of the country on cliffs and chalky soils, but in the garden it is not choosy about either place or conditions, flourishing almost anywhere, dropping its seeds at random. One needs but a single root stock to populate a garden in the second or third year.

The seed, if one wishes to harvest it, may be trapped in bags before it drops. If only the leaves are required, it is better to stop the flowering shoots from growing by cutting back; it will then produce numbers of tender light green leaves which have a faint odour of anise. If this is done the plant can even be grown in a pot or a window-box.

In ancient times fennel was already established on the European mainland. In Greece it was believed to be effective for reducing weight. Snakes were said to eat it when they cast their skins, and then to rub their faces against the stalks to strengthen their sight. Medieval man followed the snake's example, making a tea of it and applying it to the eyes on a cloth. In Rome the young shoots were cooked like asparagus and eaten as a vegetable. Charlemagne, ruler of the Holy Roman Empire (AD 800–814), promoted its cultivation throughout his dominions. In the Middle Ages it was more popular in Italy than in England but according to Shakespeare (*Henry V*) the King was fond of eating it with fish.

Up to about 1914 it was used widely as a household medicine and is a constituent of gripe water for preventing convulsions in young children. The bruised seeds were used to make tea for the prevention of flatulence, especially after eating fruit.

We might, as the Italians do, use the tender leaves in soft cheese and in salads, or put the young shoots into soups to add flavour. In some households today the fresh chopped leaves of fennel are used in the same way as parsley in the cooking of fish, and they are mingled with mashed potatoes and other vegetables in the same way as mint.

A second variety of the plant is the annual known as Florence fennel or sweet fennel (*Foeniculum dulce*). A smallish, thick-set plant, it is sown in April, bears leaves like its namesake but rather more closely packed, and its root swells into a kind of ball like an onion but with a grain like celery. It needs warmth and rich soil to be grown successfully in England and it is therefore difficult to produce a satisfactory crop in our gardens. But no visitor to France or Italy should miss sampling this 'round celery with a faint flavouring of anise' which is a common vegetable often served with butter or sauce.

Its cultivation here need not be completely fruitless however, for even if the root does not swell into the expected ball, the stems may be earthed up, blanched and cooked.

Fennel has a flavour rather like aniseed. If the young leaves are used the result is a mild flavour, while a much stronger flavour may be obtained by using the seed. Experiment alone can determine exactly how much is needed.

Mackerel with Fennel and Gooseberries

Place four mackerel, gutted and cleaned, in a saucepan with a sprig of fennel. Cover with boiling water and simmer for about ten minutes.

Cook 1 lb green gooseberries gently with 2 teaspoons of sugar for about 15 minutes or until soft. Rub the fruit through a sieve to make a sauce. Boil 2 or 3 fennel leaves for two minutes, then chop and add to the sauce.

Drain and dry the mackerel, place on a hot dish, cover with the sauce and serve.

Fennel Sauces for Fish

1 Combine 1 to 2 tablespoons chopped fresh fennel with ¼ pint of mayonnaise just before serving.

2 Make a roux with 1 oz flour and 1 oz butter. Remove from heat. Stir in until smooth 1 pint of fish or vegetable stock. Return to the heat and boil gently for 5 minutes, stirring continuously. Mix in 1 egg yolk, 1 teaspoon sugar, 1½ tablespoons lemon juice, 4 tablespoons chopped fresh fennel, salt and pepper. Return to heat and cook without boiling until sauce has thickened.

Flax

Linum usitatissimum Annual

Whoever would dream of growing flax in a small garden? Yet, ponder the word *usitatissimum* – most widely used of all – and think of all that is derived from this plant. 'Where on earth,' says Pliny, 'are there greater marvels than this?' Today we take linen, linseed oil, animal food and medicine from it. The Romans used it in addition for making cord and sailcloth, caulking their boats, making wicks for their lamps, as well as for a variety of other things.

That it is even today a universal commodity is hard to believe on seeing the graceful but timid little plant, growing on a single stalk with its sparse leaflets and pale blue blossoms. For its appearance alone it is worth a trial. A few seeds may be sown in March or April in good moist, rich soil, preferably in a spot a little removed from the rest of the herbs, where its gentle beauty may be most appreciated.

Not that it will be much use to the gardener apart from this, but surely beauty is in itself health-giving! It is only the commercial grower who pulls up the stalks of the flax plant before the leaves fall, to be despatched for linen manufacture, or takes the seeds when the flowers have died for producing linseed oil.

Foxglove

Digitalis purpurea Biennial

This is not the garden foxglove, which is a perennial and comes in many colours, but the wild variety which grows in woods and lanes in England, and which, because of its attraction for bees and its spikes of bell-shaped flowers, has been imported into the herb garden. The flowers vary from purple to rose-colour on the outside and are paler inside with dark spots.

Like many other widely-seeding plants, it is easy to cultivate, producing the tiniest of seeds which can be raised in boxes round about May and pricked out for flowering in the following year. From that time onwards the main problem for the gardener will be control, for seedlings will spring up in all parts of the garden, though the plant flourishes best in lighter soils with a little damp and shade.

Elves' gloves, ladies' thimbles, foxes' gloves, witches' fingers, dogs' fingers, fairy caps – these are only a few of the popular names given to the plant. Because of its poisonous properties it was called in Ireland dead man's fingers and red fingers, and in Scotland bloody fingers. The German name *Fingerhut* has the same meaning as the Latin *digitalis*, from *digitabulum*, a thimble.

Poisonous herbs are often used in medicine, and of these the foxglove is one of the most important. It was given in small doses in the Middle Ages for heart and kidney complaints, for dropsy and was used for binding on wounds and ulcers. In 1650 it was introduced into the London Pharmacopoeia. In 1785 an English doctor, William Withering, gave details of two hundred cases, principally of dropsy, in which he had used it to treat patients, and from then on its use became more regularly accepted. The drug digitalin, extracted from its leaves, has a beneficial action on the heart, acting as a tonic and causing the pulse to become slower and more regular, thus facilitating recovery.

Owing to its poisonous nature, it should be handled with care in the garden and kept away from children.

A second variety, *Digitalis lanata*, with a pale flower, which grows wild in Eastern Europe, is more widely used today than the English variety for the extraction of the drug.

Its healing properties are, however, the province of the medical profession and not of the layman, who should confine himself to growing and admiring it.

Garlic
Allium sativum

Garlic

Allium sativum Biennial/perennial

In Great Britain we go for the gentler onion, the leek and the mild chive. Our garlic is bought in little bundles from supermarkets and greengrocers or incorporated into salt. On the Continent, however, garlic is a vital element in cookery and has been since the days of ancient Egypt.

This is the clove garlic, the Latin *Allium sativum* or cultivated bulbs as opposed to the British crow garlic whose bean-sized bulblets and green spikes can be such a nuisance once they get into a garden. Cloves of the cultivated plant may be grown from seed or obtained from a reputable nursery, and planted in spring just like onions. As they grow larger, the numerous bulblets round the central one form new 'cloves', as they are called, for planting the following year.

The plant has a long history. It was eaten by the labourers of Khufu (or Cheops) when they built the Great Pyramid about 2,700 BC and by the Israelites during their long captivity fifteen hundred years later (Numbers XI). Soldiers, sailors and peasants of Greece and Rome valued it as an invigorating food, especially for those who had to work in the hot sun, and as a safeguard against infection. Armies on the march were never without it. For this reason it came to be associated with the rabble who chewed it raw and offended the nice sensibilities of their betters. Arthur Dent, a great Elizabethan preacher to whom Bunyan owed the idea for the *Pilgrim's Progress*, warned his hearers against 'flatterers who will speak you well though they would eat your heart with garlic'. We have heard it said that if a person rubs garlic into the soles of his feet it can within a short time be smelt in his breath. We have not tried it – yet!

Garlic pills can be bought from chemists and herbalists. A clove of garlic may be crushed and the juice diluted with water to make a tea for treating hypertension and catarrh of the intestines. It helps to cure worms and stimulates the kidneys. When used in cooking it is a useful aid to digestion.

It provides an excellent flavouring for soups and salads and is particularly good in lamb and pork dishes. For salads it is enough simply to rub a clove round the inside of the salad bowl before filling it.

Beef with Garlic and Rosemary

Add a level teaspoon of dried rosemary, salt and pepper to 2 oz flour. Cube 1½ lb stewing steak, roll the pieces of meat in the flour until completely covered. Fry in a small amount of hot fat or oil in the bottom of a pan for 3 minutes.

Place the meat in an oven-proof dish with a clove of garlic or one teaspoon of garlic salt. Add one or two onions pierced with cloves. Cover with stock or water and cook in a medium oven for 2–2½ hours. Serve with rice or baked potatoes.

Good King Henry or Lincolnshire Spinach

Chenopodium bonus-henricus Perennial

Though some people call it a weed, the Lincolnshire men knew better, and no vegetable garden should be without it, especially since it will flourish in any neglected odd corner and grows willingly, however ill-treated.

All that is needed is a single root. Planted in early spring when the soil is not too dry this will lengthen and thicken, going deep into the ground and putting up several leafy shoots with thick arrow-shaped leaves, green on one side and pale, almost white, with raised veins on the reverse. At the end of each shoot and at the leaf junctions appear the flower-spikes, each with tiny yellowish florets bunched closely round the stem and tapering to a point. By the autumn, the plant will have seeded.

The important parts are the leaves. These may be collected throughout the growing season and are at their best between April and June. Boiled, they make an excellent green vegetable. Moreover, they tend to appear before many of the usual garden crops are ready and, if the bed is large enough and the flower-spikes are cut, will continue to give leaves until the onset of winter. For this reason it is an excellent stop-gap. It is probably not better-known first, because some effort is needed to gather enough leaves for a meal, and second, because the leaves wilt quickly and are therefore not suitable for sale in shops.

When plucked the leaves feel hard and rough. On washing they turn the water a light brown, but on cooking they make a palatable soft dark—green dish rich in iron and vitamins, delightful with butter. If given an application of rich dressing – manure or humus – while growing, the stems of Good King Henry will thicken and can be blanched and treated in the same way as asparagus.

All summer long each plant drops hundreds of minute seeds, and these, although they take a long time to germinate, tend eventually to sprout small arrow-shaped leaves in all sorts of odd places from which they may easily be transplanted at a distance of about eighteen inches apart.

Good King Henry
Chenopodium bonus-henricus

Good King Henry probably got its name (*Gute Heinrich*) in Germany to distinguish it from a similar but poisonous plant (*Böse Heinrich*) known as Bad King Henry. The name, which has no connection with any king, probably refers to the wood-spirit or elf known in Germany as Heinz or Heinrich.

It has little reputation as a medicinal herb though one of its names, smearwort, indicates that it was once used to make an ointment and the leaves were used as a poultice to heal sores and ulcers.

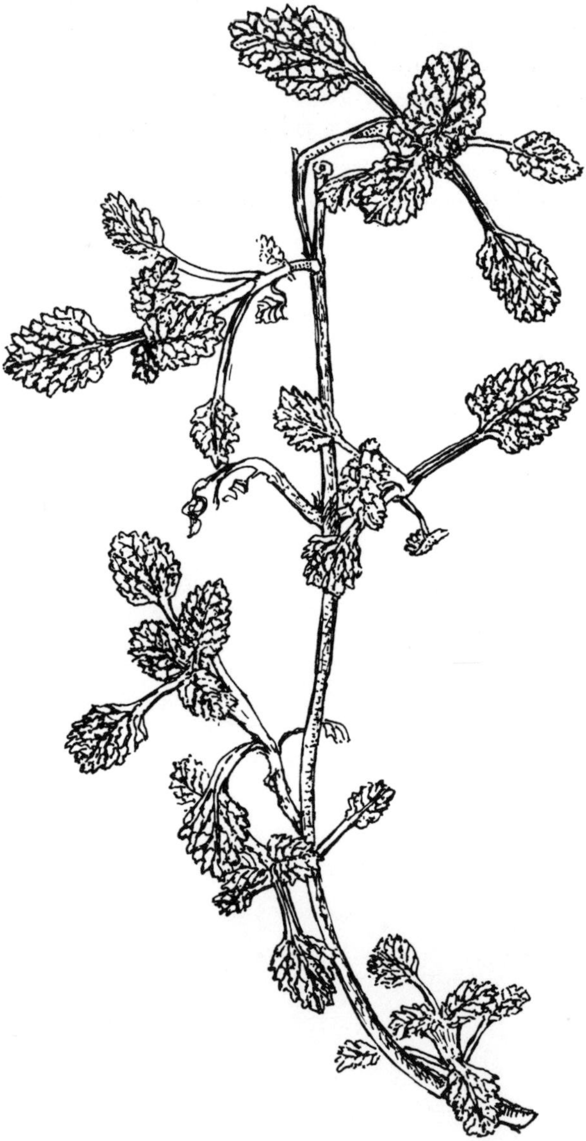

White Horehound
Marrubium vulgare

White Horehound

Marrubium vulgare Perennial

In Old English it was called *hoarhune* – the white plant – while the botanical name is said to be derived from the Hebrew *marrob*, a bitter juice. These names give a fair idea of the nature of the plant. It is about a foot high, with greyish-white wrinkled foliage, the stems coming up untidily from a single root bearing leaves in pairs up a stalk, with pale whitish flowers in small clusters at the leaf junctions, similar to those of the stinging nettle.

It grows wild in most parts of the British Isles and was brought into the medieval herb garden mainly because of its medicinal uses. It grows readily from seed sown in spring or, if not many plants are needed, from cuttings or root division. The divided plants will bloom the first season but seedlings not until they are two years old.

The flowers of this herb, however, are not needed so much as its leaves. It was one of the bitter herbs that the Israelites ate at the Feast of the Passover (Numbers IX). Its value in medieval times was in the juice pressed or infused from the leaves. Gerard recommended it for 'those that have drunk poyson or have been bitten of serpents', and for the bites of mad dogs.

Its principal value has been and still is as a medicine for coughs, colds and lung troubles, to expel phlegm, especially in the case of old and asthmatic persons. It may be taken either as a tea from fresh or dried leaves or as a syrup by adding sugar to the juice of the fresh herb and boiling it down with water. It will, if enough sugar is added, cool, thicken and form a candy. Many horehound preparations of this kind can today be bought from the chemist.

A second variety, black horehound (*Marrubium nigrum*, *Ballota nigra*), is a similar plant but has dull green leaves and purple or occasionally white flowers. It is both unattractive in appearance and has a most offensive smell. Though it was once used to prevent muscular spasm, to expel worms and as a stimulant, this practice has long been abandoned.

Horseradish

Armoracia rusticana Perennial

The horseradish is easily identified by its leaves which at the rootstock are up to a foot long and six inches in breadth, with deeply scalloped or notched edges, while the stem leaves are long and narrow, the lower ones deeply toothed. Rising to a height of three feet, the main stalk bears small white flowers on short stems.

It has been cultivated since ancient times, at first for its value as a medicine. Introduced into England in the thirteenth century, it was grated and used as an alternative to mustard. It has ever since that time been one of the main features of the English dining table and is used to flavour pickles and mustard.

Though it grows wild it can be cultivated in the garden by means of root cuttings set deeply in well-manured ground about fifteen inches apart. The roots, which are the usable parts of the plants, can be left in the soil if need be but must not be exposed to frost. Otherwise they can be lifted and kept for a time in dry sand.

Horseradish acts as a stimulant to the appetite and helps digestion. Formerly it was used as a medicine for inflammation of the kidneys and bladder. For external use a plaster can be made from the grated root for the relief of rheumatic pains.

Horseradish Sauce

Clean and grate a root of horseradish; add a pint of milk and 1 oz breadcrumbs to it.

Cook slowly for 15–20 minutes; add ½ oz of butter and salt, pepper, mustard and vinegar to taste.

The sauce may be kept for a time in the refrigerator.

A more extravagant sauce may be made by mixing 3–4 tablespoons of grated horseradish with ¼ pint sour cream, 1–2 teaspoons of wine vinegar and salt and pepper.

Houseleek

Sempervivum tectorum Perennial

It is as well to know how some of the most accessible and the commonest plants can often prove useful in time of need. Consider the well-known houseleek which, once it has taken root, puts out a circle of young growths, each a miniature of itself, all round its large green rosette. Each of these has a stalk up to six inches long, just ready to be pulled off and planted in another place to start a new generation.

Its botanical name – the ever-living roof-plant – reminds us how often we may have seen it nestling in the thatch or between the slabs of a cottage roof, in any place into which a mere cupful of loose earth may have blown. Our common name – home-plant – is exactly what the Saxons, who introduced it to England, called it. Elsewhere in Mediterranean countries it was in ancient times believed to be sacred to Jupiter. Its name in Italy is still *barba di Giove* and in France *joubarbe* – Jove's beard.

For a minor accident in the home, the houseleek growing by the door would be the first remedy on hand; the juice was used either as a poultice or, mixed with lard, as a salve to put on an inflammation or a burn. It is still given by herbalists for swellings and inflammations in any part of the body. It is said that leaves added to a foot bath and the skin of a leaf tied on the foot as a bandage and left on during the night will cure corns.

Many of these old-time remedies have long since been neglected or forgotten.

Hyssop

Hyssopus officinalis Perennial

The commonest of the garden hyssops is the blue-flowered variety, a favourite of bees, though there are others with white and occasionally pinkish blooms. It is a handsome little plant growing to a height of up to two feet with several stems rising from one root and bearing in June and July dark-green

Hyssop
Hyssopus officinalis

spearshaped leaves and flower-bearing spikes both from main and side shoots.

It is a native of southern Europe and its original name *azob* (Greek *hyssopos*) is associated with holiness. It is mentioned several times in the Bible. In medieval times, because of its pungent aroma, it was strewed sparingly with other fragrant herbs on the floors of great halls.

A hardy, sun-loving plant, it is at home in most soils and will flourish under reasonable conditions. It can best be propagated by taking cuttings or fragments split off from rooted plants and put into the soil in spring. Seeds sown in April in shallow open ground will readily produce and can be pricked out and finally placed about eighteen inches apart. One plant will last several years and is easy to renew.

It is high time that hyssop, long since given up as a kitchen herb, was again used more widely. It has a rather strong, refreshing scent with a somewhat minty, bitter taste and should certainly be one of the plants from which the herb gatherer nips off the occasional tender leaf for the salad bowl. A pinch or two adds flavour if sprinkled over fruit in a pie before cooking, and should be considered as a seasoning for broths and soups. Since it is an evergreen there is no need to worry about drying.

In continental countries the flowers and tops of hyssop are used in making sausages and the monks of La Grande Chartreuse use it in the manufacture of their famous liqueur. In the United States especially, it is an ingredient of cocktails.

The flowers and green tips were formerly and are still in some parts of the country used for making hyssop tea, regarded as a sovereign aid in bronchial, catarrhal and asthmatic conditions and for the weak stomach, especially if there is difficulty in digesting fatty substances. One is even recommended to rub the herb into poultry or the roast before cooking.

Altogether this is a useful and pleasant herb to have in the garden.

Lavender
Lavandula spica

Lavender

Lavandula spica Perennial

> *Here's flowers for you;*
> *Hot lavender, mints, savory, marjoram.*
>
> (*The Winter's Tale*)

This is the only mention of the plant in all the works of Shakespeare, for it did not come to the fore in England as a useful herb until about 1560, a few years before he was born.

There are few people who would not recognize lavender either by its appearance or its aroma. Old English lavender varies from the *Lavandula vera* from which it was developed to the many dwarf varieties for the small garden which appear in nurserymen's catalogues. A lavender hedge is a most attractive sight throughout the year, but especially when it turns purple or blue in high summer.

It is easily propagated, almost always by cuttings, for it is very rare even with the dwarf varieties for seedlings to spring up on their own. These cuttings should be taken in spring by breaking off a young shoot about four inches long, together with the small heel by which it is attached to the larger branch, and inserting this to about half its length either in a garden bed or in a pot of sandy soil. Most cuttings will take root here if cared for and will be ready for planting out in the following year: four feet apart for larger varieties, two feet apart for the dwarf lavenders. In this way renewal is easy and should be done every few years as the plants, especially the larger ones, tend to become straggly after three or four seasons. They can of course be trimmed but care should be taken not to cut off too many of the larger shoots at one time.

Lavo, from which the botanical name of the plant comes, is Latin for 'I wash'. The dried flowers were used by the Romans for scenting the bath, and it is believed that they introduced it into Britain. In the Middle Ages it was to be found in monastery and herbal gardens. Once adopted in the mid-sixteenth century its popularity as a household favourite grew rapidly and it was cried in the London streets in Stuart times, and indeed, lavender sellers are familiar figures even today.

Today we use lavender exclusively for scent. The flowers are taken by cutting off the spikes just as they open, and these spikes are usually shortened and then laid out on clean trays or sheets away from direct sunlight. When properly dry the flowers can be rubbed off, sifted, put into sachets for scenting the wardrobe, into potpourri jars, or stored for later use. Oil is obtained from the fresh flowers by distillation on a commercial scale and may be bought at stores. At one time lavender water, prepared at home from the flowers, was used as a medicine and as a tonic, while lavender oil was used as an application for sprains, rheumatism, toothache and neuralgia. They have both now been largely superseded by other preparations.

Lemon Verbena

Lippia citriodora Perennial shrub

Originally found in Chile, this shrub was introduced into Britain about two hundred years ago and is now established in many gardens not only because of its characteristic odour but also because of its convenient size, which never exceeds fifteen feet in southernmost localities, and for the beauty of its foliage and its many small pale purple flowers which bloom in August.

No other herb except lemon balm has this scent, which is specially suitable for the potpourri jar because it is so long-lasting.

It has no particular medicinal value although an infusion of the leaves has been used like those of mint to give gentle relief in cases of flatulence and indigestion. A few leaves, fresh or dried, make a pleasant addition to China tea.

Lemon Verbena
Lippia citriodora

Lovage
Levisticum officinale

Lovage

Levisticum officinale Perennial

Has any poet even mentioned this plant in his writings? Have any legends or myths accumulated round it? In spite of its attractive name, we believe not.

It may be that it is too overbearing, too greedy for space to be so favoured, for one small root sporting a couple of celery-like leaves will, in the space of a year or more, develop into a giant, pushing its shoots with their dark green leaves out from the base, slowly smothering smaller and more weakly growths within a few feet of it. It is a mercy that, unlike tansy, its roots do not creep out underground and throw up more giants like itself.

Some gardening books tell us that its umbels with their greenish-yellow flowers and brown seeds rise on their stalks to a height of four to five feet. Our own reaches seven or even eight feet or more, its shape resembling the flowering stalks of some gigantic cow-parsley. No wonder Culpeper says in his *Herball* that one plant only may be more than enough.

All the same, if there is room in the herb garden it should be given a trial. Planting a small rootlet in spring is enough. It will quickly establish itself, live through any climatic crisis and can be divided for propagation.

Lovage was a well-known plant in the old English garden though its Latin name indicates its probable original home – Liguria in northern Italy. It was valued in olden days for its odour and flavour, strong, like very strong celery: should one wish to have celery flavour in sauces, soups, stocks and stews without the celery, the juice of lovage is the answer. Its leaves when young are tender and can be used sparingly in salads, but when they grow large, as most do in a plant such as this, they become hard, leathery and useless to the cook.

Its medicinal value in the Middle Ages was high. The roots and fruits were used, like those of angelica, for flatulence and loss of appetite, and the leaves as an infusion for jaundice, gravel and other urinary complaints.

The ideal place for it is in a vacant corner of the garden or

Common Mallow
Malva sylvestris

well to the rear of the herb bed, but it is far too good a plant to discard out of hand.

Mallow

Malva sylvestris – Common Mallow
Althaea officinalis – Marsh Mallow Perennial

The common or blue mallow is the one which gives greater satisfaction to the gardener. It is a large plant growing to some three feet in height with bright flowers ranging from purple to pink. When in bloom in June the plant is at its best but becomes less attractive as the flowers die off and the leaves rapidly lose their colour. It is best propagated by the splitting of roots.

For value as a source of medicine the marsh mallow is preferred. This grows wild in southern England, especially in damp meadows, near ditches and on salt marshes from Lincolnshire southwards, and in many parts of Europe. It can also be cultivated. Its serrate, deeply-veined leaves and delicate pale purple flowers grow on erect spikes topped by numerous buds. It will grow from seed but is better propagated by cuttings or root division in autumn. As its name indicates it flourishes best where moisture is present.

While not among the principal medicinal herbs the marsh mallow has always had a reputation for healing properties. The Greeks called it *althea* – the healer. Pliny maintained that a person taking it in the morning need fear no disease that day. It contains mucilage and the root, peeled, cut into shavings, boiled and allowed to stand in the water, then drunk with some sweetening agent, is a tonic for the intestines and kidneys. Tea made from the leaf is used in the treatment of respiratory troubles and for soothing sores in the mouth. Both roots and leaves used for poulticing are good for reducing inflammation, and marsh mallow ointment was a famous old-time remedy for all sorts of minor injuries in the home.

Marigold
Calendula officinalis

Marigold

Calendula officinalis Annual

Nothing is more eye-catching than the golden expanse of a long, wide bed of double marigolds in a public park or an ornamental garden – not the French or African variety, but the pot marigold, the cultured descendant of the yellow star found in the old English garden. The golden flowers can be raised from seed, and will seed themselves in their turn. The little single flower which Shakespeare knew, 'that goes to bed wi' the sun', is a beautiful addition to any garden.

Cultivation presents no difficulty whatever. Throw down a few seeds in April; scrape the soil over them and in a very short time the light green leaves shoot up and need thinning to about a foot apart. As they flower, cut off the dying heads and the plant goes on and on until it exhausts itself at the end of the season and dies. Hence its name *Calendula* – the plant which is blooming in some place or other throughout the whole calendar year.

Today we grow it, admire it, and that is all. Our forefathers knew differently. They pulled the petals from the flowers and dried them, put them into broth, soaked them in sugar water, candied them 'to strengthen and comfort the heart', steeped them in water to bathe their tired feet, used the yellow dye which so readily comes off on the fingers to colour cheese and to bring out the gold in their hair. They used the petals to sprinkle sparingly on green salad and to make the marigold tea which they found so soothing for intestinal troubles.

From the leaves and flowers a tincture is made by dissolving in alcohol, and this is counted a sovereign remedy for wounds, carbuncles, sprains and stings. 'No suppuration,' says one doctor, 'seems to be able to live in its presence.' In the American Civil War calendula tincture, compresses and ointment were among the principal agents for treating wounds. The tincture and ointment may still be bought, but a few plants in the garden, used as our ancestors used them, are just as good.

Marigold Buns

- 2 eggs
- 3 oz sugar
- 4 oz plain flour
- 2 tablespoons fresh or dried marigold petals

Separate the eggs. Add the sugar to the yolks and beat well. Gently mix in the flour and the marigold petals. Beat egg whites until stiff and add to the yolk and flour mixture.

Put in greased bun tins (about 20), top with marigold petals and a sprinkling of sugar. Bake for 20 minutes in a moderate oven.

Marigold Yeasty Buns

- 6 oz plain flour
- salt
- $\frac{1}{2}$ oz fresh yeast (or 2 teaspoons dried yeast)
- $1\frac{1}{2}$ oz sugar
- 1 egg
- 2–3 tablespoons milk
- chopped marigold petals

In a warm bowl mix the flour with a pinch of salt. Make a well in the centre. Cream the yeast with 2 tablespoons lukewarm water. Put yeast into the well and mix with a little flour. Cover with a cloth until the yeast has doubled in size. Then work in the remaining flour, adding the sugar, egg and milk.

Knead until smooth. Form into buns and place on a greased tray. Brush over with milk. Sprinkle with chopped marigold petals and cook in a moderate oven for about 30 minutes.

Marigold Rice

Sauté an onion in fat, add $\frac{1}{2}$ lb long grain rice and fry gently for a few minutes. Pour on 1 pint vegetable stock in which marigold petals have been soaking, add a few pinches of rosemary and a pinch of salt. Cover and cook until the rice is tender.

Sprinkle with grated cheese and dot with a little butter. This is a delicious dish which makes an unusual change from potatoes.

Marjoram

Origanum onites Perennial

In ancient Greece a herb grew on the sunny slopes which gave a bright purple stippling to the grey and green of the hillsides. People called it 'joy of the hill', the herb of happiness. They appreciated its fragrance, crowned young betrothed couples with wreaths made of its stalks and believed that if it sprang up on a grave it would assure the well-being of the departed.

This was marjoram, a small herb with spreading roots which throws up numerous stalks to a height of about a foot, and these bear fragrant purple flowers. We call it pot marjoram (*Origanum onites*) because, if planted in pots or window-boxes, it is capable of living through the winter. It can be grown from seed and planted out in the open in April or May and here it will spread, throwing out its runners which will root and if broken off form new plants. It might even, if not watched, stray from the border and grow into the edge of the lawn. Though living through the winter, the plant becomes dormant, to take on new life in the spring.

A frailer plant is sweet majoram (*Origanum marjorana*), an annual which is best propagated by seeds sown in heat in March or in the open in late April in dry, light soil and transplanted about a foot apart. They germinate slowly, usually throwing up single shoots bearing lighter grey leaves than the perennial kind, with small clusters of tiny white flowers – hence the plant's popular name of knotted marjoram. It has a more delicate, sweeter smell than other varieties.

Marjoram leaves may be used fresh from the garden for salads and, with others such as basil, thyme and savory, for seasoning and stuffings. Cut and stripped from their stalks when they are at their best, just before the flowering period in July, they are easy to dry and keep their aroma and flavour for a long period. Like chervil, marjoram forms part of the *bouquet garni* so important in French cooking.

The variety mainly used in medicine is the wild marjoram (*Origanum vulgare*) which grows in Britain and in various parts

Pot Marjoram
Origanum onites

of Europe, North Africa and Asia. It makes a sweet-tasting, rather oily tea, considered a good tonic for nervous headaches and impaired digestion.

Marjoram Liver Dumplings

½ lb calf's or pig's liver
2 oz butter
grated rind of a lemon
½ teaspoon fresh or dried marjoram, chopped
1 clove of garlic, crushed
4 oz self-raising flour
stock or broth
salt and pepper

Simmer the liver in water for 2 minutes, then remove all fibres and mince finely. Blend the butter with the lemon rind, marjoram, garlic, salt and pepper and add the liver to this mixture. Gradually add the flour until dumplings can be formed, and simmer them in stock or broth for 10 minutes.

Mint

Mentha Perennial

Green or garden mint (*Mentha viridis*) and spearmint (*Mentha spicata*) are the best-known names of the common mint grown in the English kitchen garden. They are so near to each other in appearance and qualities that what difference there is between them hardly matters.

There must be few people who have not seen a bed of mint in spring, a brilliant green with its shoots about a foot high bearing their purple flowers in the leaf junctions. Some may have grown it and learnt to cut away the underground creeping stems or to hem them in with deeply-sunk tiles to prevent the plants thrusting their roots into some neighbouring bed.

Due to its strong rapid growth mint can, if not watched, become a nuisance, pushing between border stones, making it necessary to remove them and pull out the fragile white stems before they can do more harm. But vigilance brings reward in the shape of a tidy garden and a nice harvest of dried leaves to garnish green peas, new or mashed potatoes, to make mint sauce

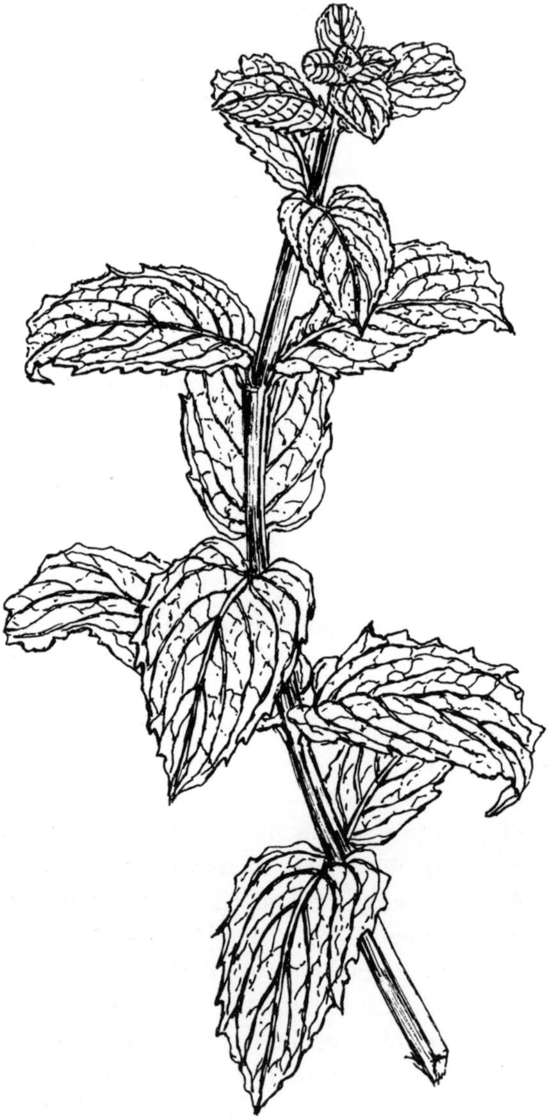

Apple Mint
Mentha rotundifolia

for roast lamb or fragrant sweetened mint tea. Cultivation is easy, even in a window-box, but to prevent deterioration a complete replanting in open ground is advisable every three or four years. There are so many varieties of the original mint that a complete list is impossible in this context. Some gardeners prefer one, some another. Bowles mint is a larger variety with pale, hairy leaves, and serves exactly the same purpose as green mint. Orange or Eau de Cologne mint has larger, rounder and darker leaves than the garden variety and is best used for teas or for mixing with fruit drinks, possibly with borage if the garden provides it. Pineapple mint, used for the same purpose, has smaller, darker leaves, spreads less rapidly and is not quite as hardy. Care must be taken to keep other soil-greedy plants out of its way or it will succumb. Apple mint (*Mentha rotundifolia*), as its name indicates, has its own characteristic flavour. These and others can be obtained from herb farms.

Peppermint (*Mentha piperita*) deserves separate mention, for it is the source of most mint medicines. Cultivation differs in no way from that of garden mint, though it is not much grown in the domestic garden, most people relying on the well-known peppermint oil which is on sale in most chemists and grocers. A few drops of this added to water or given on a lump of sugar are used to allay dyspepsia. Peppermint tea, sometimes with the addition of elderflowers, 1 oz to a pint, allowed to stand about seven or eight minutes after scalding, is an old remedy at the onset of a cold.

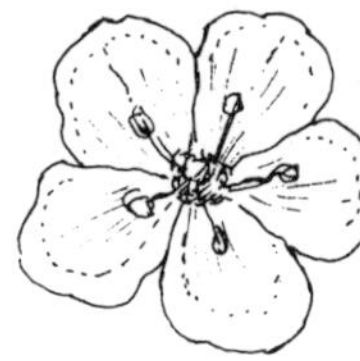

Mullein
Verbascum thapsiforme

Mullein

Verbascum thapsiforme Biennial

The tiny seeds of the Great Mullein drop in odd places all over the garden. In the first season it puts up a little rosette of grey flannel-like leaves; in the second the one thick stem which rises from its centre, grows and grows much like the hollyhock until it reaches a height of up to ten feet, with clusters of buds round it on all sides, which burst open into bright yellow blooms at irregular intervals. Often the stem puts out smaller secondary stems which do the same, making the plant look like a large many-branched candle among whose tiny lights the bees hover ceaselessly.

Cultivation is no problem. Once a mullein has grown in the garden its tiny progeny continue to appear, sometimes remaining dormant for long periods before bursting into life perhaps a year or two after having been dropped, the little grey leaves showing themselves on the soil surface at any time during the growing season.

No herb in the whole list has more varied local names as this, for it is compared to wool, flannel, rags, velvet, blankets, fluff, felt, a candle, taper, torch, rod, staff and club among other things.

Nor is any herb more conspicuously British for, though it grows in other parts of Europe, Asia and the New World it has nowhere else been regarded with so much respect and even superstition. Its dried leaves, used as tinder, gave it the name of candlewick plant or Our Lady's Taper and, dipped in fat, it was used as a torch at funerals and on other solemn occasions.

Besides being an ornamental plant the mullein is an old household remedy for the derangement of one main function – that of the lungs and bronchial tubes. Farmers gave it to their cattle as a cough medicine and therefore it earned the name of bullock's lungroot. To make a household medicine the young clean leaves were boiled for ten minutes in a pint of milk, then strained to take out the hairs and given warm with or without sugar for a cough. Its taste is most unlike the other medicines for the same complaint, being bland, rather unusual and not in the least

Parsley
Petroselinum crispum

unpleasant. Alone or with other herbs it can be used as a tea for all manner of chest complaints.

Though it no longer has a place in the British Pharmacopoeia, homoeopathic physicians employ the tincture of the leaves for hoarse, dry night coughs, and in dilutions for the involuntary escape of urine.

Parsley

Petroselinum crispum
Petroselinum sativum Biennial

Parsley is a common herb, and the bright green crimply-leaved variety (*Petroselinum crispum*) is best known in the English garden, though there are others, especially the broad-leaved variety, known as the Hamburg parsley (*Petroselinum sativum*), cultivated mainly for its thick roots which may be cooked and eaten like the parsnip. In both varieties the leaves grow directly from the root on longish brittle stems. When using them for any culinary purpose the younger leaves should be chosen.

Parsley starts life as an obstinate little plant, taking a long time to germinate, sometimes up to eight weeks. It can be sown as early as February in a box or pot and can then be transplanted into a bed in April or May leaving up to a foot between plants according to their kind. There are two ways to encourage germination in the open – either to pour boiling water on to the soil immediately before sowing or, after a few fine days when the soil has had the chance to warm up, to sow in shallow drills, sprinkle soil over the seeds and then water using a fine rose, with water just hot enough to put the hand in. In a small garden one need not be too severe about thinning.

Seeds sown in August will, under normal conditions, provide leaves very early in the following year. Being a biennial the plant produces leaves only during the first growing season, and in the second year the flower stems begin to come up. If these are removed it will, with feeding and watering, be productive into the late summer.

Though a parsley bed will give some green leaves all through a mild winter, drying is essential for a good supply. This should be done more quickly than in the case of most other herbs, preferably in the oven when it is cooling off after cooking, again and again until it is thoroughly crisp and ready for rubbing. It should then be stored in an airtight container away from the light.

The plant was dedicated by the ancient Greeks to the goddess Persephone who returned to earth from the underworld every spring and it was therefore closely associated with seasonal and human death and regeneration, being woven into wreaths for tombs. Even in some parts of Britain it is still considered bad luck to transplant it – a superstition we are only slowly discarding.

Sadly, British cooks tend to be rather unimaginative in their use of parsley, though it can be used in a variety of dishes, and it should have a much wider use for the table. It can be chopped and eaten with butter on bread or added to young dandelion leaves and mixed with yoghurt. Parsley stalks can be dried and powdered to give a green food colouring or the juice squeezed from fresh leaves can be used as a colour for cake decoration.

Medicinally parsley tea has always been recommended as being good for rheumatism, for the kidneys and bladder and for inflammation of the prostate gland. During the First World War kidney complaints arising from conditions in the trenches were treated with it.

Parsley Meat Sauce

- 2 hard-boiled eggs
- 1 teaspoon vegetable oil
- 1 tablespoon lemon juice
- 1 heaped tablespoon chopped parsley
- 2 egg whites
- salt

Remove the yolks from the hard-boiled eggs and mash them with the oil. Add a pinch of salt, lemon juice, and parsley.

Whisk the egg whites until stiff and fold into the yolk mixture. Mix well together with a little cream or top of the milk. Serve with lamb or pork and new potatoes.

Parsley Biscuits

Rub 4 oz margarine into 4 oz plain flour until it becomes crumbly. Add 1 tablespoon dried parsley, the juice of one small onion or a small teaspoon of garlic salt.

Mix all together with milk and water to form a stiff dough. Roll out ½ inch thick and cut into shapes. Bake in a moderate oven for 15 to 20 minutes.

These are delicious served hot with soup or cold with cheese.

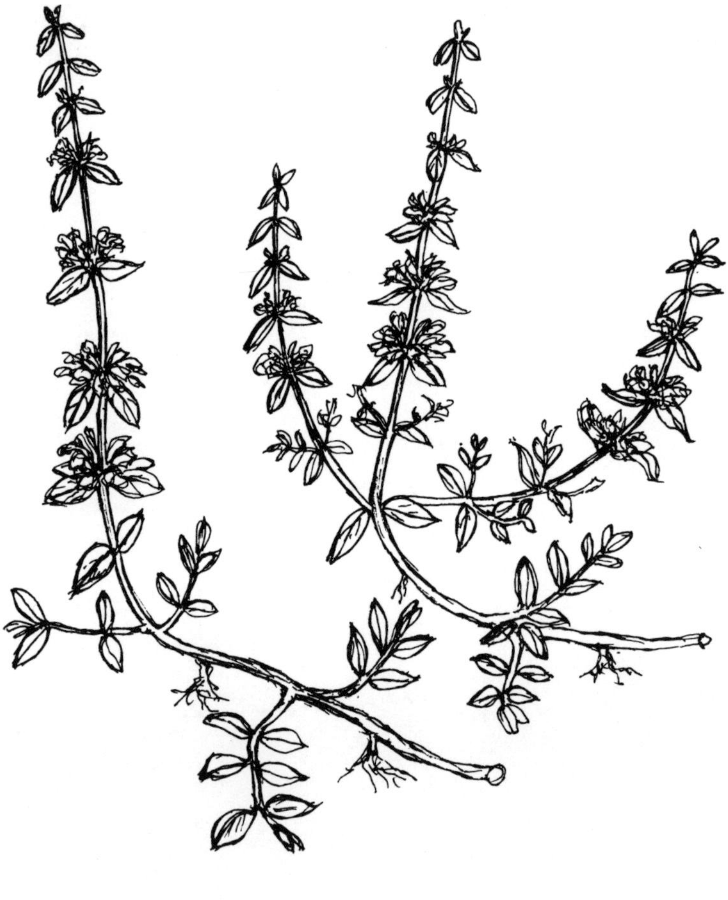

Pennyroyal
Mentha pulegium

Pennyroyal

Mentha pulegium Perennial

As its botanical name indicates, this plant is really one of the mints but it is treated separately here because in the first place the best-known variety, unlike the mints, is a creeping plant and occupies a different place in the herb garden; and second because its odour and taste are different, more pungent and much less agreeable than any of the others. It would not, even in the smallest quantity, be of much use in the kitchen.

It is a hardy plant which grows wild all over the world, especially near streams and in moist soil. Before London became heavily built up it was common to find it among the ponds and marshes where the Mile End and Whitechapel Roads are now, and out in the western suburbs. Its leaves grow in pairs along the creeping stems, which in July and August bear small clusters of purple blooms. For this reason it is an ideal plant for the edges of paths and between stones, but it needs control or it will spread into places where it is not wanted.

There is no problem about propagation. A small rootlet will soon throw out several stems which, being on ground level, will easily take root if unhindered and can be cut off and planted elsewhere.

The reason for its great popularity in old-time markets is that, in the absence of more modern medicines, it could be made into a tea which when sweetened was by no means unpleasant and to which lavender or rosemary flowers could be added. It was an old-fashioned remedy for coughs, bronchitis, indigestion and for cleansing the blood. It is not long since mothers whose daughters were facing the first onset of menstruation always kept dried pennyroyal handy in the kitchen cupboard for making a soothing tea. In earlier days a sprig or two was sometimes kept between the bedclothes to repel fleas, and hung in bedrooms because it was believed to cleanse the air, or made into garlands as a remedy for headache.

Today we use it entirely as an ornamental creeping plant. Like chamomile it is pleasant to walk on and, providing our noses are not too near, adds nicely to the scent of the herb garden.

Lesser Periwinkle
Vinca minor

Periwinkle

Vinca major – Greater Periwinkle
Vinca minor – Lesser Periwinkle Perennial

The greater and lesser periwinkle are well-known wild plants and have been brought into the garden mainly for their tendency to spread and form attractive ground cover on what would otherwise be unsightly vacant spaces. The Swedish botanist Linnaeus (1707–1778), who made the first effective classification of plants, gave it the name *vinca* from the Latin *vincio*, I bind, because of the profusion of intertwining shoots which creep along the ground, easily take root, and continually make new plants. The lesser periwinkle, whose leaves and flowers are both smaller, is often recommended in gardening catalogues for this.

Because of its green foliage which is retained throughout the winter and its many attractive purplish-blue flowers borne in May and June, it has always had a special place in the garden. In medieval times it was customary to pluck it in certain phases of the moon to assure prosperity and protection from sickness and possession by devils. The Italians call it *centocchio* – the hundred-eyes – and Chaucer included the 'pervinke' in the floral garment of Cupid.

It has no place in the kitchen, but both greater and lesser varieties have been used medically as mild purgatives and to stop bleeding. Homoeopathic preparations have been prescribed for haemorrhages and the tincture is still applied to wounds. In a dilution of a hundred to one, it is given internally for milk-crust (eczema of the scalp) in children.

Purslane

Portulacca oleracea Annual

Though not a native of Britain, purslane can readily be grown here. It is a small herb with fleshy red stems and thick dark green leaves, and tends to spread itself on the ground. Its small yellow flowers appear in July.

It is not seen much in England nor does it figure much in French cookery, though it is one of the ingredients of the French soup *bonne femme*. It is, however, useful as an addition to the spring salad bowl when the leaves are at their best, and the shoots may be used as a pot-herb. Drying is not advised, but the thicker shoots can be picked for winter usc.

It is grown from seed sown under cloches in April and in the open ground in May. It likes sandy soil and sun, but especially when young, it should be watered frequently, pouring in the water a little distance away from the plants and letting it percolate down to the roots. If the flowers are nipped off when in bud and the plant cut back after removing leaves it will last longer. It is advisable to try a first sowing and, if it flourishes and is appreciated, to repeat every month.

Though purslane was praised by old-time herbalists as an external application for sores, for bathing the eyes and internally for worms in children and the retention of urine, it no longer figures as a medicinal herb.

Raspberry

Rubus idaeus Bush

Hardly a herb in the botanical sense of the word, the raspberry is included here primarily because of its reputation in this country as a household medicine.

It grows wild in many parts of Great Britain and Europe, has a bright-green biennial stem with fine spines and perennial roots which spread quickly underground. Once the shoots have flowered in May and June and fruited the following month they are cut to the ground so that the younger ones may flourish and bear fruit in the following year.

The uses of the fruit for dessert, confectionery and wine are well known. Medicinally the most important part is the leaf. Dried, powdered and mixed with the powdered bark of slippery elm it makes a poultice for cleansing and healing wounds. An ounce infused in a pint of boiling water has been used from time immemorial as a medicine during pregnancy and, taken warm at frequent intervals, sweetened if necessary, is still considered to promote quicker and easier labour.

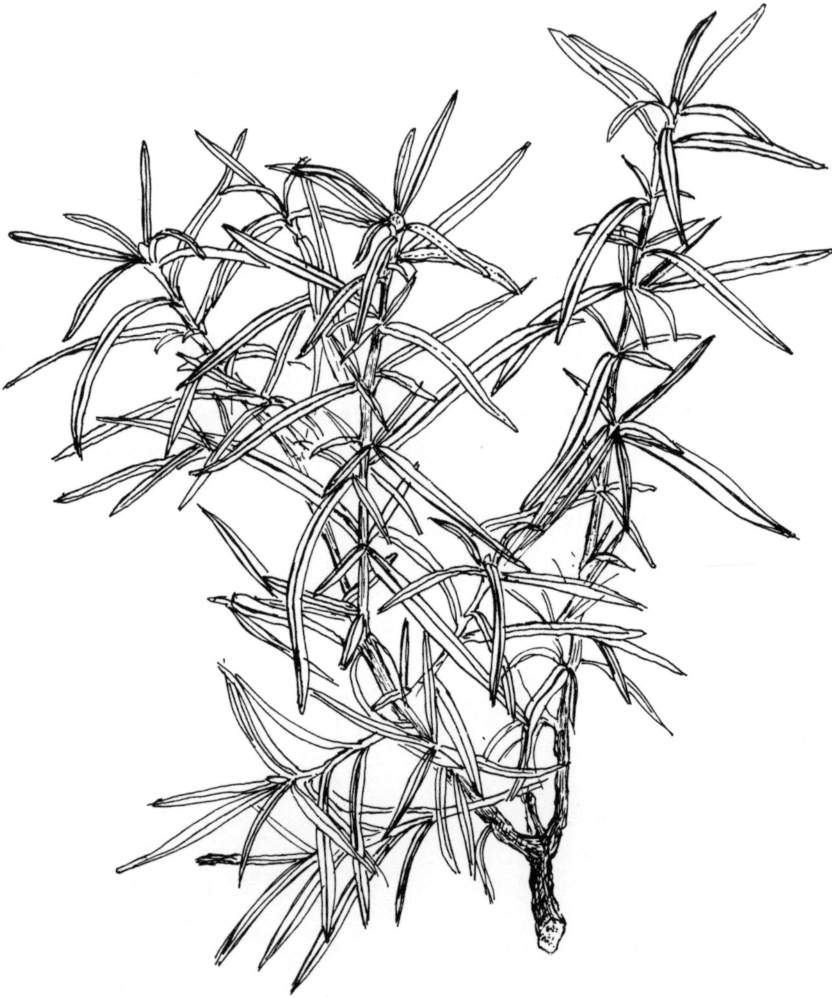

Rosemary
Rosmarinus officinalis

Rosemary

Rosmarinus officinalis Perennial

This 'queen of the herb garden' is an evergreen bush growing to a height of some five feet. It has upright or almost upright branches, each bearing many small dark-green leaves along its length about an inch long and resembling large pine needles. Its tiny mauve flowers cling close to the branches among the leaves in spring, but the bush is apt to surprise the gardener, especially in the southern counties, by producing a wonderful display of flowers at other times during the year, even around Christmas in mild winters.

A gentle, well-behaved bush, rosemary grows best on well-drained soil in a sheltered position, and needs little care or trimming. It can be propagated by seed, but an easier method is to take small cuttings in pots between June and August, keeping these in a shady spot where they will readily take root and can later be planted in individual pots or in the open. For gifts around Christmas, a number of rosemary plants, each in its own small decorated pot, can be very handy.

The most pronounced feature of rosemary is its scent, and no herb garden should be without it. One needs only to touch its leaves to carry the scent on the fingers for some time. It is very attractive to bees and the leaves, collected after flowering has ended, may form the basis of a potpourri jar.

Throughout the ages rosemary has been the herb of remembrance and was once used by lovers as a symbol of fidelity. It was wound among the flowers in a bride's coronet, and its dumb language, according to Sir Thomas More, made it the chosen emblem at funerals and in burial grounds. In the Middle Ages it was burnt in sick rooms as it was believed that it could prevent infection. Visitors took it into prison as a precaution against the spread of gaol fever (typhus). In time of plague it was one of the herbs carried in the hollow gold head of the physician's walking-cane.

Oil of rosemary is used as a hair lotion, and to make a rinse which is particularly good for dark hair, boiling water may be poured over both leaves and flowers, strained and used when lukewarm. It is one of the ingredients of Eau de Cologne.

Medicinally it does not possess the virtues of hyssop or comfrey but an infusion of its leaves may be drunk as a tea to relieve headaches, and as a tonic. The leaves, put into white wine and allowed to stand for some days, make a tonic drink. A stronger infusion may be obtained by chopping up a tablespoonful of leaves, leaving them in a half a pint of cider or white wine for some days, and then straining. Rosemary tea is made from tops and leaves – two teaspoons to a pint of boiling water, sweetened with sugar or honey.

Cooking with Rosemary

Dried and powdered rosemary sprinkled over joints of beef, lamb or pork before roasting gives a piquant flavouring to the cooked meat.

Lamb or pork chops with a layer of onion and a sprinkling of rosemary on them, wrapped in foil, put into a tin and cooked in a moderate oven for 30 minutes make a delicious meal.

A plain biscuit mixture becomes something special if half a teaspoon of rosemary and some chopped walnuts are added before placing in balls on a baking sheet. Flatten with a fork before cooking at 350°F or gas mark 4 for 25 minutes or until light brown.

Rue

Ruta graveolens Perennial

Rue is an evergreen plant of Mediterranean origin introduced into Britain by the Romans, since which time it has been a prominent and handsome feature of the herb garden.

A hardy plant, about eighteen inches high, it will grow almost anywhere, even in the most barren soil, and can best be propagated either by cuttings taken in spring or by rooted slips. It has a woody stem which lives through the winter to give it something of the character of a shrub, with bluish-green leaves and bright yellow flowers coming out between June and September.

Rue
Ruta graveolens

Probably because of its strong, disagreeable scent it has throughout the ages been associated with sorrow, repentance and bitterness.

In his *Discourse on Sallets* John Evelyn wrote that the ancient Greeks used it to clear the sight, that their artists ate it in large quantities, and that the Italians put it into their salads. In the Middle Ages branches of rue were used to sprinkle the holy water in church. Shakespeare refers to it as 'the sour herb of grace' (*Richard III*). It was regarded as a precaution against infectious diseases, so that judges were presented with it, along with other herbs, at the beginning of their sessions to ward off gaol fever, and physicians carried it on their visits to patients during epidemics of the bubonic plague.

It has figured in heraldry. The thistle and rue (And-rew) are interlaced in the collar of the Order of the Thistle. It was included on the coat of arms of Albert, the Prince Consort through his descent from the kings of Saxony, and of his son Edward Prince of Wales (later King Edward VII).

This plant is, when undiluted, a skin irritant, but in small quantities it is used in baths for the treatment of rashes and other skin diseases. The same rule regarding quantity applied to the tea which used to be made from the dried and powdered leaf. Otherwise it is not to be recommended, especially during pregnancy. The homoeopathic physician uses tincture of rue diluted with water as a lotion for bruises, especially when the bone is affected, for rheumatism and sprains of the wrist and ankle, and internally for weak-sightedness, for over-exertion and for piles, especially soon after the onset.

Saffron

Crocus sativus Perennial

No English herbal can be considered complete without at least a mention of this noted herb, now fallen almost into disuse but once essential for industry because of its yellow dye and in the home because of its unique flavour and its medicinal powers.

It is a low plant growing from a bulb or corm very much like the ornamental spring crocus save that the saffron corm is planted in July in good, light, preferably chalky soil, and the large lily-shaped reddish-purple flowers appear in September. The only parts of the plant of any value are the stigma and part of the style which were picked out and dried to make the pressed cakes or bars of the herb that were sold in London markets centuries ago.

It has had a romantic history. It appears to have been exported from Kashmir and Persia to the Near East and China in ancient times, to have fallen out of cultivation and to have been rediscovered during the twelfth and thirteenth centuries in Moorish Spain. Naturally the Moors, wishing to keep the secret of their precious commodity, forbade export of the plant but, according to legend, a pilgrim smuggled a single corm in his hollow staff and brought it to England. The stretch of chalky soil in north Essex was found especially suitable for its cultivation and the town of Walden, the centre of the industry, was henceforward known as Saffron Walden.

Apart from its traditional reputation for helping to make people light-hearted it was in former days valued for curing flatulence, inducing perspiration, bringing on periods in women and checking haemorrhages. Today we see little of it, but to some the taste of a good Cornish saffron cake is still something of a treat.

One warning: this plant is not to be confused with the poisonous autumn crocus (*Colchicum autumnale*) which is found wild in meadows. This plant is also used medicinally in minute doses for treating cases of gout, but should never be used by the layman.

Sage
Salvia officinalis

Sage

Salvia officinalis Perennial

He that would live for aye
Must eat sage in May.

The Latin *salveo* – I am in good health – and the name *salvia*, the sage plant, are derived from the same root-word, which proves that to some extent sage was valued as a health-giving herb in ancient times.

Its original home was in the Mediterranean lands, whence it was speedily carried to northern Europe and later on to North America.

It is a stout and hardy plant growing to a height of two feet or more, its woody stalks having long greyish-brown leaves in pairs, each leaf finely marked with veins. In late July and August the stalks of the common sage are topped with a profusion of mauve flower-spikes. The leaf, pressed between thumb and forefinger, has a strong, bitter and lasting odour unlike that of any other growing herb except perhaps clary, which is of the same family.

Cultivation is no problem. Cuttings taken in spring will strike readily, but it is easier to take rooted pieces from an old plant and these can be obtained in greater numbers if needed by earthing up the soil round the roots of the large plant and allowing rootlets to form. The plants should be replaced at least every three or four years as in time they are liable to grow leggy and misshapen. This can be held off a little during the first two years by trimming long growths in spring.

The leaves, which can be cut in June and September, are the most used part of the plant. Great care should be taken in drying, and, once dry and rubbed, the herb should be kept away from all moisture or it tends to gather a musty smell and becomes useless.

To most people the phrases 'sage and onions' and 'sage stuffing' are familiar. It is also used in large quantities for certain kinds of sausages, soups and gravies, especially on the Continent.

Sage tea once enjoyed enormous popularity, and was the best-known drink apart from ale and mead among English peasants before the importation of Ceylon and China tea. The Chinese even preferred it to their own, which they exchanged for it at the rate of two or more pounds to one of dried sage. Its reputation as a cleanser of the system and a blood purifier has lasted until modern times, and it is a component of many herbal tea mixtures. Its flavour can be modified by adding lemon juice and rind, or basil, sweetened with sugar or honey.

Besides common sage there are three other main varieties. The broad-leaved sage is less hardy and does not bloom in England. Because all its energy goes into the formation of leaves it has a greater reputation as a kitchen herb than the common sage. Red sage, with purplish-red leaves, is said to be a better gargle for sore throats, and the variegated sage is grown mainly as an ornamental plant.

St John's Wort

Hypericum perforatum Perennial

How revealing some of these botanical names are! *Hypericum*, the plant which flourishes below the heath-land; *perforatum*, having translucent spots on its leaves which are really small oil-glands.

This is one of the wild herbs native to the temperate regions of the Old World which, because of its medicinal value, especially in the Middle Ages, was introduced into the herb garden. It grows on an erect, smooth stem rising to a height of some eighteen to thirty inches, bears small offshoots with pale-green oblong leaves and from June to August clusters of bright yellow flowers marked with darker dots and lines. Crush these and dark red smears stain the fingers from the glands at their base. Bloodstains? Those who believed in the Doctrine of Signatures thought so, and concluded that it was a signal herb for the curing of wounds.

They were right, but not for that reason. One would like to think that 'this thrice blessed herb' for the relief of man, is connected with St John of Jerusalem after whom was named the

Order of Knights Hospitallers. This may or may not be, but before modern vulneraries came into being it was universally applied to wounds from sword and spear, especially where the wounded part was rich in nerves. Even today few herbal ointments used for ulceration, cuts and bruises do not contain the extract from this plant. Taken internally, an infusion of St John's wort speeds up the bodily functions, stimulates the appetite and 'with his floures and seed boyled and drunken, provoketh urine and is right good against the stone in the bladder' (Gerard). It is used by the homoeopath in very high dilution for nervous depression following injury, for the effects of shock and to speed up the healing process.

One other variety, *Hypericum androaemum*, a wild, shrubby herb more resistant to cultivation, found in similar places to St John's wort, has large oval leaves and yellow flowers. In medieval times it earned the name *tutsan* (French *toute saine* – all-heal) for similar virtues.

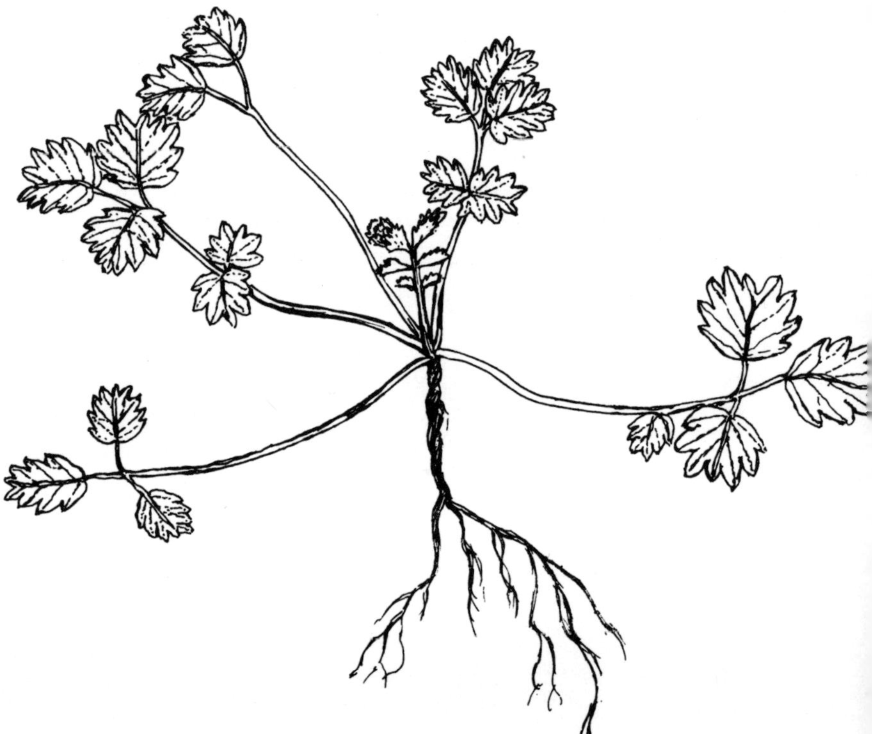

Salad Burnet
Sanguisorba minor

Salad Burnet

Sanguisorba minor Perennial

A small perennial, rarely exceeding a foot in height, the salad burnet came to be cultivated in medieval gardens because of its use in salads and its medicinal virtues. Like borage, it has a smooth cucumber-like flavour and was used to make drinks cool and mild.

'It has two little leaves like unto the wings of birds, standing out as the bird setteth her wings out when she intendeth to fly.' (Turner, *New Herball*, 1551.) In its wild state it grows mainly on chalky soils in southern England and makes excellent feed for sheep and cattle. The leaves, which are a good additive to salads, should be picked when young, and their growth is assisted if the plant is not allowed to flower.

Cultivation is easy, for the plant remains green throughout the winter. Seed sown in March will produce young plants and these can be put out later, leaving nine inches to a foot between plants. Alternatively, roots can be split, but this is rarely necessary as salad burnet is a self-seeding herb.

It has long been collected in the wild as a medicinal plant and taken as a tea is useful for skin ailments, promoting digestion and helping the function of the kidneys.

Savory

Satureja montana – Winter Savory Perennial
Satureja hortensis – Summer Savory Annual

The distinction between these two herbs is clearly seen from their botanical names – *montana* meaning 'of the mountain', and *hortensis* 'of the garden'. The first is a hardy perennial with almost white flowers, a penetrating scent and a sharp flavour, the second a more delicate annual with an altogether sweeter scent.

Winter savory is a small bush with scores of separate shoots growing from its rootstock. These bear tiny dark-green leaves and in July the whole plant turns a whitish colour as if sprinkled with wood ash. This is caused by the thousands of tiny flowers it bears. Like lavender, it is one of the best bee plants in the garden

Winter Savory
Satureja montana

and an ornament to any flower bed. It will grow in almost any kind of soil, weathers the winter well and, if the shoots which have flowered are regularly cut back, will last for years.

Summer savory is grown from seed sown under glass or indoors in March or in open ground in April. Each seed sends up a single shoot and, when these are large enough, they may be planted out about six inches apart where they will grow into miniature bushes about a foot high bearing small, narrow spear-shaped leaves of a lighter green than the winter variety. The flowers, which appear in July, are purple. This is the herb which is collected wild in Mediterranean regions, mainly for drying and use in cooking. It could, like sweet marjoram and basil, be tried for indoor cultivation as a pot plant in summer.

Being of Mediterranean origin, it was well known to the Greeks who considered it as sacred to the satyrs, the goat-footed woodland deities, hence its name *satureja*. The Romans used it as we use tarragon and mint, to flavour vinegar and sauces. It is useful, mixed with other herbs, for seasoning, and imparts a nice flavour when added to cheese, scones or biscuits.

Both kinds of savory dry well and keep their flavour after freezing, but they are not easy to rub, especially the winter variety, as the thin stalks tend to get mixed up with the leaves. Once dried they should be kept in an airtight jar so as not to attract moisture.

Savory has no uses today as a medicinal herb though the juice was once used diluted with water for eye-drops. A sprig of either winter or summer savory rubbed on a bee sting will quickly bring relief.

Used sparingly in the kitchen it is a good flavouring for salads and cooked vegetables, especially green peas and broad beans. If added to cabbage, sprouts and other leaf vegetables it helps to prevent the smell which arises from cooking.

Solomon's Seal
Polygonatum multiflorum

Solomon's Seal

Polygonatum multiflorum Perennial

One need know little about herbs to recognize this garden plant, a favourite, especially in the south of England. Originally a native of wooded land, it is not often found wild in England and has been brought into the garden because of one single virtue – the beauty of its delicate arching stalk with its small cream-coloured, lily-like bell-flowers.

Once planted, especially in a shady position, it needs little care, for its creeping rootstock rapidly spreads and throws up more and more shoots – hence the need for control. All that is necessary to form a new plant is a small portion of the rootstock of an established one, taken up and transplanted, preferably in autumn.

The origin of its uncommon name is disputed. Some believe it arises from the small round scars on the root which resemble the stamp of a seal, others because the plant was good for 'sealing' or healing fresh wounds and broken bones when used as a poultice.

The virtue of the plant lies in its root. Powdered, it was once applied to wounds, piles and ulcers and was considered more effective than a piece of beef for putting over a black eye. Made into a tisane it was used for soothing irritation of the urinary passages, and during menstruation.

English Sorrel
Rumex acetosa

Sorrel

Rumex acetosa – English Sorrel
Rumex scutatus – French Sorrel Perennial

The more common of these two plants is the English sorrel which grows wild in fields and meadows all over the country as well as in most parts of Europe and North America. Standing some two feet high with spikes of small reddish-green flowers in June and July, it is easily recognized among the neighbouring grasses and herbs. The leaves are spear-shaped and up to six inches long where they cluster round the base of the plant on longish stalks. Because of its value as a vegetable and an additive to cooked food it has long been found in herb gardens.

It is easily grown in a fairly damp situation either by sowing or by splitting off from a larger root, and needs growing space of about a foot each way.

The French variety has a more succulent, fleshy and crisp leaf than the English and may be distinguished by the leaf shape, as it has two spreading lobes at the base like those of Good King Henry. Both varieties can be eaten as a vegetable but some people do not take to it owing to its rather sour taste, which is due to the oxalic acid in the plant. This makes it unsuitable for people who have a tendency to rheumatism or gout. As a sauce or a dressing for lamb, veal, pork and poultry or put into salads and sauces it once had a great reputation in England, but is now used much more in France, especially the French variety.

In former times it was recommended as a remedy for scurvy as it is rich in vitamin C, but now that other sources of the vitamin have been found, it is not used at all as a medicine.

Southernwood or **Lad's Love**
Artemisia abrotanum

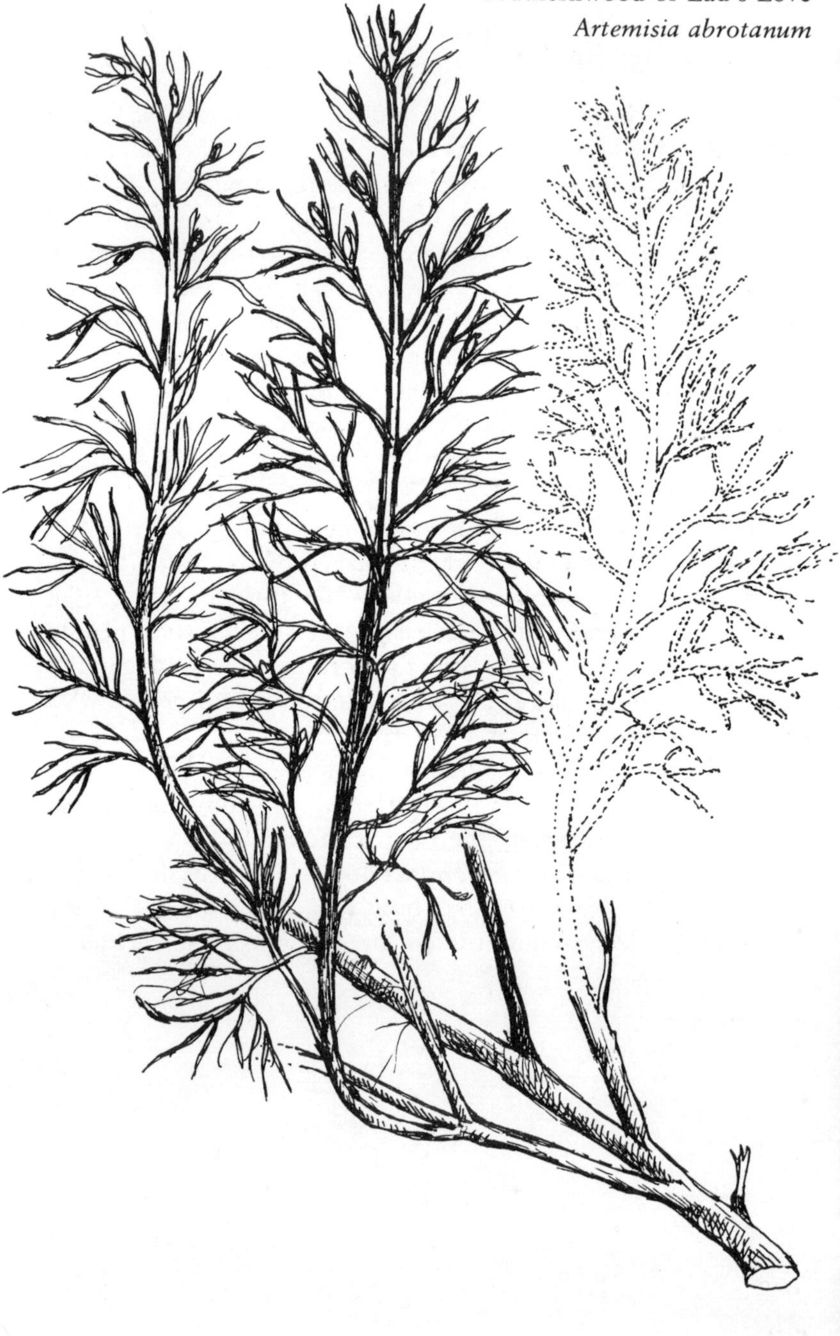

Southernwood or **Lad's Love**

Artemisia abrotanum Perennial

Few old-time gardens are without this fragrant herb. As its name indicates, it grows wild in Mediterranean countries but here in Great Britain we rarely or never see its flowers. It was introduced into this country in the mid-sixteenth century.

It has a place today for no other reason than its keen aromatic scent and, if it is cared for, its good looks in summer. It produces long, woody stems bearing light-green feathery leaves with tufts at the top where the shoot grows ever higher. Cuttings taken from the plant either in spring or autumn will root readily and, when the fine weather comes, grow rapidly, ultimately attaining a height of some three feet or more.

The plant is not particular as to soil or climate and can be put in any odd spot in the garden. For appearance's sake it is probably better placed out of the direct line of vision because in winter the feathery leaves still cling to the long erect branches or else fall off, leaving only the gaunt grey-brown stalk, which is probably why it acquired the name Old Man. Happily it takes kindly to being trimmed and can thus be made to look a little more tidy.

Southernwood was a favourite herb of our ancestors; it is best known as lad's love in country districts because it was once part of the bouquet which young men presented to girls. Its keen scent was believed to prevent infection and it was placed in drawers and wardrobes to keep moths away. The herb was once used as an infusion for gastric disorders and was powdered and given in treacle to children to expel worms. Today it is no longer used as a medicine.

Sunflower
Helianthus annuus

Sunflower

Helianthus annuus Annual

One cannot but marvel at the stupendous powers of growth of the garden sunflower. A small seed put into well-manured ground in April will make a small rosette of leaves in May and then thrust a single rough stem upwards, sometimes to a height of twelve feet, its bud bursting into an enormous disc composed of hundreds of small florets surrounded by a fringe of bright yellow petals, the whole reminding one of the sun in total eclipse with its bubbling corona of flame. No wonder the Incas of Peru marvelled at its power of growth, used it to make crowns for their priestesses and represented it in gold.

In Britain it is cultivated as a garden flower, and when the bloom has died it is thrown away. In Russia, the seeds are saved and eaten in much the same way as we eat sweets. However, we are learning, and today sunflower seeds are highly recommended for their nutritional value and vitamin content. They are hulled, put in bags and sold by the pound, mainly in health food shops.

Together with cereal crops and the soya bean, the sunflower is an increasingly important agricultural product, both because of its value as a foodcrop and also for its oil.

Sweet Cicely
Myrrhis odorata

Sweet Cicely

Myrrhis odorata Perennial

It is hard to understand why sweet cicely has been condemned by some as being of little valuc and not worthy of a place inside the garden fence, as it is one of the greenest and most beautiful of herbs, especially in the spring. It may be because it is land-hungry, more so even than lovage, creating large thick roots and throwing out runners in all directions. For this reason it may be unwelcome in a very small garden but it is worth trying out, because it is such an attractive plant.

Its beauty lies in its large lacy leaf, whitish and downy when it first comes up, then spreading out from the base and all up the thick stem, bursting at the top into more bright-green fronds and finally large compound umbels of white blossoms to which bees are attracted. When bruised, leaves and stem have the odour and flavour of sweet anise. The largish dark-brown fruit rather like onion seed retains its odour and it is said that beeswax, scented by maceration with sweet cicely, was once used to freshen and polish oaken floors. In due time the seeds are broadcast about the herb bed where they germinate easily, and the seedlings can be transplanted so that, should an old plant be destroyed, there will always be another to take its place.

Sweet cicely could be used more in cooking than it is. The young leaves or cold boiled roots are a tasty addition to salad, and in some places the roots are used as a vegetable. By adding a few young leaves to sour fruit, such as gooseberries, when stewing, sugar can be saved. Sweet cicely has also been used to make medicines and tonics, especially for young girls. Boiled and eaten with oil and vinegar, it can be a comfort and a source of energy for old people.

Some of its many popular names – British myrrh, anise, sweet chervil, shepherd's needle, sweet fern, and cow chervil – are apt to make people confuse it with other plants at first but, once having grown and appreciated it, they will never do so again.

Tansy
Tanacetum vulgare

Tansy

Tanacetum vulgare Perennial

In the vegetable sections of their catalogues some nurserymen include information about a most effective insect repellent – tansy, grown from seed and planted out when six inches tall, one plant to four dwarf or runner beans. Most forms of ant, they claim, will not appear when tansy is present.

If there is anything immortal in a garden it is tansy. Once established it will send out underground runners to great distances and if not watched they will get mixed up with the roots of other plants, so beware! What is more, it prospers throughout the rainiest season and the most severe drought will not kill it. In the long hot summer of 1976 we had plants in only six inches of soil with concrete underneath. These we watered not more than twice, and in spite of all the drought, small bunches of dark green leaves were still growing in September at the end of its stems, their odour as persistent and keen as their hold on life.

All the same, it is worth having a plant, if only for the rich dark-green leaves, segmented and toothed, which grow at its rootstock and up its stems, and for the masses of yellow buttons with no petals which appear in August and September. Both are a boon to the flower-arranger.

The tiniest pinch of the young leaf chopped very finely lends a tang to a salad, and that is all that can be said of its present-day use. In olden times tansy pudding was eaten at Easter after six weeks' fasting because it was thought to clear out the system, and it was one of the strewing herbs because of its disinfectant properties. Tansy tea was given in moderate doses to sufferers from gout and to children to expel worms. But too much or too strong a solution is dangerous, causing abdominal congestion, so as a culinary herb this beautiful plant is best left alone.

Tarragon
Artemisia dracunculus

Tarragon

Artemisia dracunculus Perennial

There are two varieties of this herb, French and Russian. The former is a native of southern Europe, grows to a height of between two and three feet and is used, in France more than anywhere else, in cookery. The Russian variety grows taller, but the inferior flavour of its leaves means that it is far less popular than the French variety.

When the plant is fully developed the root looks like an involved sort of clump which sends out small rootlets, and these can be taken off in March or April to form new plants. Cuttings can be struck in July.

In spring numbers of tiny shoots appear on this root-clump and every one grows into a stem bearing long narrow leaves. Then in August come the tiny inconspicuous flowers which are of no use whatever for seeding. The leaves have a strong, rather hot taste and a unique, rather unpleasant odour; this and its height make it unsuitable to be kept near the clothes-line. The stalks die off in winter and should be cut down to the ground.

Tarragon is dried by hanging the stalks in bunches in partial or total shade during warm, sunny days, well separated so that the air can get to them. Any damp is fatal to the flavour. Even when dry and crisp the leaves should be secured in airtight containers or they will lose what virtue is in them. It is not as easy to dry as most herbs and, in any case, dried tarragon is vastly inferior to the fresh green leaf.

The name comes from the Latin *dracunculus* – the little dragon, which it is still called in some parts of England. A French popular name is *herbe au dragon* and in Italian *dragoncello*, possibly because the leaves were once used for taking away the effects of stings, and against the bites of mad dogs.

Today its uses are entirely confined to cookery. A few leaves chopped and added to a salad add variety to the flavour. It can be incorporated into sauces and stews but is best known in England because of tarragon vinegar. This can be made at home by steeping the fresh leaves in vinegar, preferably wine or cider, just covering them, allowing them to stand for anything up to ten days, then straining and bottling. Tarragon can be blended with butter with salt added, to use as a dressing for grilled meats.

French-Style Chicken and Tarragon

Put a medium-sized chicken into a casserole dish and cover with stock, one tablespoon fresh or dried tarragon, salt and pepper. Cook in a moderate oven for about $1\frac{1}{2}$ hours.

Make a roux with 1 oz of butter and 1 oz of flour and seasoning, and pour the liquid from the chicken on to this. Pour this mixture into a saucepan and simmer gently, stirring continuously, until the sauce is thick and creamy. Pour over the chicken and serve.

Thyme

Thymus serpyllum – Wild Thyme
Thymus vulgaris – Garden Thyme Perennial

Did ever a plant of Mediterranean origin become so dear to English people as thyme? It has been bred and cultivated till there are almost enough varieties to fill a small gardening catalogue. It is luxury, comfort and necessity all at the same time, and its by-products are in evidence in soaps, scents, creams and lozenges in every chemist shop.

All varieties are cultivated forms of the wild thyme, the bee-plant mentioned by Shakespeare, which lends a special flavour to honey. Common or garden thyme is the eldest and best known of its progeny. Because it grows to a height of only six to nine inches, it makes a good edging to the garden, and flourishes in pots or rockeries. It bears large numbers of small dark green leaves and in July lilac-coloured flowerets. It very much resembles a miniature bush because its woody stems do not die back in winter.

Cultivation is not difficult. The plant can be grown from seed sown indoors in March and in an open sunny spot in April. It is not quite as easy to take rooted shoots from it as from many other herbs, as it does not throw out as many rootlets, but a longer

Wild Thyme
Thymus serpyllum

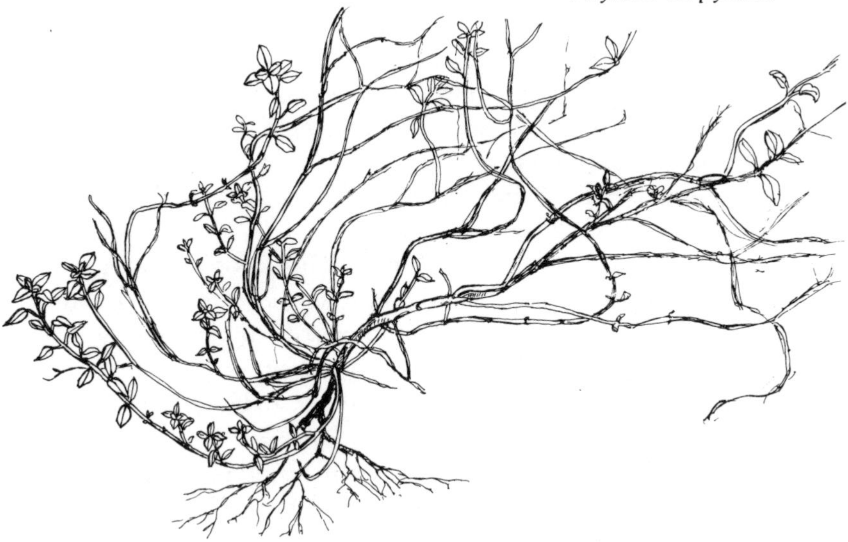

Garden Thyme
Thymus vulgaris

shoot can be made to take root by being pegged down or layered and covered with soil. If this is done about April it will within three months have rooted and can be planted out.

The name of the plant is believed to have originated in the Greek word *thumos* meaning courage, and it was long an emblem of vigour and stamina. Ladies embroidered it on the garments they presented to their knights in the Middle Ages, and it was a famous emblem of insurgents in France at the time of the Revolution.

In the kitchen it deserves, though it does not receive, as much attention as mint, for it is an important ingredient in seasoning, stuffing, sauces and stews. The flavour of thyme can be transferred to salad oil or vinegar, as can that of most other strongly-scented herbs, by allowing the leaves to soak for a period, then straining.

The herb itself is not used to a great extent in medicine except in home-made tea which is used for coughs, catarrh and digestive troubles. The oil of thyme known as *thymol* is extracted commercially and is among the best known of all herb oils as an antiseptic, a cleanser of wounds and in particular as an ingredient of many cough and throat remedies.

The many different varieties of the plant can easily be obtained from nurseries. The most popular are golden for show in the garden, lemon for its exquisite scent and flavour, and creeping thyme for the lawn and edging.

Valerian

Valeriana officinalis Perennial

'Phu! That is valerian!' wrote a medieval doctor referring to the plant's offensive smell. It was popularly known as 'Phu' for centuries.

It grows wild in damp places in England, especially near marshes and along the banks of rivers and ditches. It was introduced into the herb garden at an early stage.

The beauty of valerian is in striking contrast with its disagreeable scent. The thick rootstock sends up one sturdy shoot, a flowering stem with segmented leaves arranged in pairs which bursts at a height of some three or more feet into paired stalks, each of which bears one or more clusters of pinkish flowers. Clay soil and a damp climate suit it well and under such conditions it will seed, spread and take over a large part of the garden if not controlled.

It is difficult to say what can be done with it in a garden beyond admiring it. A tea made from its roots is repulsive to the taste; we would never dream of using it as a salad herb though the Anglo-Saxons did. The Elizabethans put it into broth and laid slices of the dried root among their clothes. It attracts cats almost as powerfully as catmint and has such a fascination for rats that the Pied Piper of Hamelin is said to have used it as a bait.

However, it does have its uses in medicine though they are hardly domestic. Oil of valerian is extracted from the roots on a commercial basis; it has a sedative effect and is good for the heart. But if taken to excess valerian produces the very symptoms that small doses of it are given to cure. We should do little more than admire this handsome herb.

Wormwood
Artemisia absinthium

Wormwood

Artemisia absinthium Perennial

Bitter herbs are valued both as tonics and in the manufacture of wines and liqueurs. As the Latin name indicates, wormwood is an important ingredient of absinthe and chartreuse and was, until 1915, when the French Government prohibited it, used in the manufacture of Pernod. The name of the aromatic white wine vermouth is a corruption of wormwood.

The plant grows abundantly over all parts of Great Britain and Europe and has now spread to America and New Zealand. The stem, which rises from a perennial root, is covered with fine silky hairs, giving both it and the deeply divided leaves a whitish appearance. From the upper part of the main stem branch several sub-stems which between June and September bear spikes of small flowers which look like tiny drooping yellow balls.

It has long been a familiar plant in the herb garden, and is propagated either by seeds which may sow themselves in autumn, or by division of roots or by cuttings in spring. The latter is preferable for the small garden. As it may reach any height between two and five feet it needs plenty of room.

Because of its bitterness it is used figuratively both in the Bible and by Shakespeare as a symbol of sorrow and anguish. Only Thomas Tusser (*c.* 1524–1580), that great apostle of British husbandry, mentions some of its real uses – to keep away fleas, to strew with rue on the floors of rooms and to be taken as a medicine 'for comfort of hart and braine'.

Today it is not used domestically, though oil is extracted from it on a commercial scale and used in medicines for aiding digestion and increasing the appetite. But even the wines and aperitifs which contain it can do harm if drunk in excess, as it contains an ingredient which affects the central nervous system.

Yarrow

Achillea millefolium Perennial

Yarrow is a weed and, although it figures in some herb garden catalogues, the grower should be aware of some of its less desirable features. For instance, it uses more than one method to survive and spread, sending out underground shoots as well as broadcasting its seeds, so that one may see yarrow plants springing up in all sorts of odd places. On a country walk one could hardly miss it, and so, if there is a lane with a grass verge near home, that may well be the place where the enthusiast can find it.

It is a tough little plant which has followed western man across the Atlantic and the Indian Oceans to North America and the antipodes. *Millefolium*, or milfoil as it is sometimes called – meaning the thousand-leaved – is a good description, for the leaves, like those of its taller relative the tansy, are so deeply cut and feathery that one alone may be divided into sixty or more segments. It bears white or pale lilac daisy-like flowers crowded in clusters on stalks about eighteen inches high.

The first part of its name *achillea* refers to the legend that Achilles, the hero of the Trojan War, was taught by the centaur Chiron to make a salve from it to heal his wounds and those of his men. It has always been used as a healing plant, for which virtue it was given many other popular names – soldier's woundwort, knight's milfoil, *herbe militaris*, carpenter's weed, bloodwort, staunchweed and sanguinary. A leaf rolled and applied to the nostrils helps to stop nosebleeds, hence its other popular name, nosebleed.

Tea made from dried yarrow leaves stimulates the flow of gastric juices, is good for kidney and other internal complaints and for excessive menstrual flow. It has a bitter and rather hot taste and may be sweetened with sugar or honey. Pinches of young leaves may be added, sparingly, to salads.

These may be some of the reasons why the cottager found it handy to have yarrow in his garden. Flower-arrangers will probably appreciate it more because the flowers, when hung to dry on their stalks, do not drop and are useful for decoration in winter.

Field, Woodland and Roadside Herbs

Agrimony

Agrimonia eupatoria

Agrimony, a herb which grows freely in grassland, may be transferred to the garden where it will spread by dropping its seeds, though this is not often done. Its common name, church steeples, refers to the spikes of bright yellow flowers that rise to a height of one to two feet. The seed vessels have stiff hairs with hooked ends which cling to the clothing of passers-by and the coats of animals like tiny puffballs, hence another popular name, the stickleburr.

It contains a high proportion of tannin and in the past it was used to help wounds to heal. Today, an infusion of the herb – stem, leaves and flowering tops – is used as a gargle for a relaxed throat, as a mouthwash, for inflammation, especially in cases of chronic skin rashes, and as a tea to improve the function of the digestive organs, especially the liver.

Betony

Betonica officinalis

The wood betony has square, hairy stems and rough, heavily-veined leaves. It grows to a height of one to two feet and from June to September bears purplish-red flowers. It is to be found mainly in woods and copses in England and in various parts of Europe. Though not normally a garden plant it may be introduced by the collection and sowing of seeds from a wild specimen.

Its virtue as a medicinal herb has been known from Roman times. A physician to the Emperor Augustus (*d.* AD 14) wrote a long treatise listing no fewer than forty-seven diseases for which it could be used. It was a favourite plant in medieval physic gardens and single specimens are still found occasionally among the ruins of old abbeys.

The whole herb, collected and dried, was considered a remedy for many ills, especially diseases of the blood, the malfunctioning of the digestive system and the liver. In taste the infusion resembles China tea and can be made more palatable by the addition of honey and lemon.

Burdock

Arctium lappa

Most people who take country walks know this tall, handsome plant which grows profusely on river banks and roadsides, especially where there is ample moisture. Its name, 'burr-dock', describes it well though it belongs not to the dock, but to the thistle family. There must be few children who have not thrown the burrs to stick to each other's clothes. The burr, which also clings to the hair of animals, is the plant's way of spreading its seeds. In England the spread can be controlled but in some vast areas of open land, such as in Australia, the burdock has become a major nuisance.

It used to be much more popular than it is today. Mineral water manufacturers, especially in the midlands and the north, still sell bottles of dandelion and burdock, a drink which has a delightful taste which somewhat resembles liquorice with a touch of aniseed. It is certainly a more wholesome drink than some of the concoctions on the market today.

The whole herb – root, leaves and fruits – has a long-standing reputation inferior only to that of sarsaparilla for purifying the blood.

Greater Celandine

Chelidonium majus

Though poisonous, the greater celandine was highly thought of in past centuries as a medicinal herb, and often found a place in the cottage garden. Standing up to two and a half feet high, it is related to the poppy, bearing small four-petalled yellow flowers which give way in late summer to long narrow seed-pods.

The bitter orange-coloured juice of the plant was once used as an eye-drop and is still put on warts to remove them. Given as an infusion it has a mildly sedative effect. The crude drug is used under medical supervision for inflammation of the bladder, gallstones and hepatitis.

Centaury

Centaurium minus

The single tall stem grows to a height of up to a foot, putting out regular pairs of pale green leaves and, towards the top, smaller branches all carrying tufts of tiny pink flowers. It is found on roadsides, on dryish pastures and on chalky and sandy soils, especially cliffs and hillsides.

Its medicinal properties were well known in ancient Greece (see page 4); it was used everywhere before the discovery of quinine against fevers and still is in some parts of Europe. One of its old English names is feverwort.

The whole herb, collected and dried in July, has a bitterness which from ancient times has given it a reputation as an appetizer and as an aid to digestion. For this reason it is a constituent of various herbal teas and drinks known as bitters which are taken before meals.

Chickweed

Stellaria media

Chickweed is about as universal as the starling or the buttercup. Wherever man's foot has trod, there it is to be found. Hardy, self-propagating, it is a nuisance to the gardener who roots it up mercilessly. It spreads rapidly during a mild winter and when pulled up, its tender stalks leave behind tenacious, many branching hair-like roots in the ground, often interwoven with those of other plants. Its tiny flower, which gives chickweed its botanical name, 'the little star', is a favourite with birds.

The tenderness of the growing stalk is one of its virtues. Left untouched it will spread and cover the ground with lush green leafage, and this can be used uncooked with lettuce in salad or boiled and eaten like spinach or Good King Henry. However, its nuisance value in the garden tends to obscure any good points it may have.

Its healing powers seem either to have been forgotten or neglected. A green ointment for treating sores and piles was once made by chopping and cooking the leaves and stalks in lard. Fresh, cleaned and put into a muslin bag, it is still an old country recipe for drawing the cores out of boils and abscesses.

Coltsfoot

Tussilago farfara

This is a proud little plant that never asks for the best. Walk along a railway embankment, through an old ash tip or a barren stretch of shale or even slag, and there you may well find it pushing up its delicate grey stalk, muffled all over with pointed pinkish-tipped bracts as if to protect it from the cold, for the coltsfoot blooms in February even before its large hoof-shaped leaves appear. The stalk is topped by a many-petalled yellow flower shaped like the brush of a chimney-sweep.

Its old name was *filius ante patrem* – the son (the bloom) before the father (the leaf). In some northern industrial towns on whose vacant lots it flourishes it is called a cleat, elsewhere ass's foot or bull's foot.

Tussilago means cough-destroyer, and this is its principal use. An ounce of the leaves in a quart of water boiled down to a pint, mixed with liquorice and sweetened, helps to do the trick. Coltsfoot rock, available from most herbalists' shops, is a more

Coltsfoot
Tussilago farfara

palatable remedy for the young. Combined with other scented herbs its dried leaves are smoked without any of the ill effects left by tar and nicotine. It was in great demand during the last war when tobacco became scarce.

The leaves are also used externally in compresses for varicose veins and ulcers.

Cowslip

Primula veris

A field of cowslips today is a rare sight, but centuries ago it was one of the commonest of wild flowers in England. Shakespeare, who was no botanist, mentions it seven times in his plays, more than any other flower.

Its botanical name, meaning the true primula, denotes the time of its flowering – early spring. Its tightly-packed leaves radiate from a single root, and the flowers, each umbel rising on its own stalk no more than about six inches high, bear the small yellow bells for which it is so well known. It has been suggested that its uncommon name is derived from the Old English word *leac* which means a plant, cows*leek* having been corrupted to cows*lip*.

It was a favourite in earlier days for making homely remedies, chiefly cowslip wine and cowslip water, both from the flowers alone. The juice of the flowers, an ointment made from them or white wine in which the flowers had been allowed to stand – all these were used as applications to remove wrinkles. Cowslip water and wine were recommended for relieving restlessness and insomnia. Today the plant has no medicinal uses.

Dandelion

Taraxacum officinale

Some years ago a rather shy young man on a visit to a French family fell into conversation with a very prim elderly lady, and the conversation came round to the subject of gardens.

'Ah oui, je suis jardinier,' he said, then tried to explain that his

Dandelion
Taraxcum officinale

garden was overrun with '*dent-de-lion*'. The old lady was puzzled.

'*Dent-de-lion? Qu'est-ce que c'est?*'

In the end he had to describe the plant – a leaf like a lion's tooth and a many-petalled yellow flower on a hollow stalk. The old lady understood and her face lit up.

'Ah-h-h-h! *Pissenlit!*'

This quality of the dandelion for stimulating the bladder has inspired the French from time immemorial to scour the roadsides and fields to cull the fresh green central leaves of the dandelion rosette to put into their salads. It is one of the most beneficial herbal aids to the functions of the gall bladder and kidneys, partly because, unlike other remedies, it is not in the least poisonous, and partly because though bitter it has no perceptibly nasty taste when mingled with other herbs. Only recently have its virtues as a salad herb been generally appreciated in this country but now nurseries are retailing special seeds for sowing in the garden, with instructions to cover and blanch the leaves like chicory.

The main medicinal value of the plant lies in the root which has long been used medicinally, and has been powdered for the making of a coffee substitute. The flowers with sugar, oranges and raisins make a pleasant wine.

Eyebright

Euphrasia officinalis

Here is a plant that cannot possibly be given a place to itself in a garden, for it revels in company. It even takes some nourishment, in addition to what it gets from the soil, by entwining its own roots with those of other plants, so that it is to some extent a parasite. Naturally it is among grasses that it is most at home. Being an annual plant and not more than about eight inches high, its mild parasitism does no apparent harm to any of its neighbours. Its small bright-green serrated leaves and, from July onwards, its tiny purple and white flowers, may be seen among grass, especially in meadows and chalky soil.

Its popular names, in French *casse-lunette* (smash-spectacles) and in German *Augentröst* (eye-comfort), allude to the power it has to reduce inflammation and to deal with eye-infections. The parts used are the flowers and the foliage from which a lotion is made with water for an eye-bath. An infusion of eye-bright has long been used in digestive and gall-bladder disorders. With coltsfoot, betony and other herbs it is a constituent of herbal tobacco.

Feverfew

Chrysanthemum parthenium

A native of southern Europe and Asia Minor, the feverfew has long been familiar in Britain and its branching stalks of yellow daisy-like flowers are a common sight in our fields and hedge-rows. It has been taken into many a herb garden where it quickly establishes itself and multiplies through casting its seed. It is a perennial which can also be increased by cuttings or division of roots.

Its name is a corruption of *febrifuge* – the dispeller of fevers – and this is what the bitter infusion was primarily used for. In days when ague (malaria) was present in marshy districts people used to drink the tea and bind the spikes of the plant round their wrists as a preventative. Taken internally it served to ease tension and pain during menstruation and acted as a general tonic.

Groundsel

Senecio vulgaris

This plant should be mentioned if only to say that, with chick-weed, it is a permanent garden pest, especially where the soil is mainly clay, for it revels in damp conditions.

It is easy enough to pull out of the soil but it is virtually im-possible to eradicate the plant completely before the quick-developing downy seed-heads disperse far and wide. It was given its botanical name *senecio* because of the appearance of these seed-heads which on the plant resemble the hair of an old man (*senex*). The English word groundsel, from the Saxon *grunde-*

swelge, ground-swallower, alludes to its rapid power to spread, for if one plant is left undisturbed, it is said that it is capable of producing a million more in a single year. As with chickweed, a mild winter does not check its development, and the spring often finds autumn-dug ground a rich green with a mingling of both weeds covering it.

The animal kingdom gives it a mixed reception. Goats and pigs like it; horses and sheep avoid it. The only people who welcome it are the owners of rabbits and birds who feed the shoots to their pets.

Medicinally it is not used today though it was once taken as an emetic and an infusion of the plant was used as a wash to soften the roughness of chapped hands.

Juniper

Juniperus communis

A long-living evergreen shrub, the size of the juniper depends largely on the conditions under which it grows. On the chalky high land in southern England it is a small tree; in the Scottish highlands it is diminutive and may even hug the ground. Related to the pine, the juniper produces small spiky leaves and green berries which turn a dark blue as they slowly mature. They are the part of the plant from which the extracts – juice and oil of juniper – are obtained.

It is common knowledge that the juniper berry is the chief flavouring agent in gin, which derives its name from the plant. The dried berries are used as an infusion for stomach troubles and for stimulating the function of the kidneys and bladder, but juniper in any form should not be used except under medical supervision.

Lady's Mantle and **Parsley Piert**

Alchemilla vulgaris and *Aphanes arvensis*

This small plant, standing not much more than a foot high, earned the name *alchemilla* because in the Middle Ages the alchemists believed that the dew which gathered on its leaves in the morning would help them to transmute base metals into gold. It grows in wet grassland, on the verges of woods and near the summits of mountains where moisture is plentiful. It is widespread in Asia and North America as well as in Europe. Its pale green leaf which, held downwards, looks like a miniature cloak, has given it the name. It bears small green flowers in loose clusters on a branching stem.

The lady's mantle was used three centuries ago to prevent miscarriage, but a smaller variety, known as parsley piert, is used as a remedy nowadays for its action on stones in the bladder. For this reason it was given the name 'parsley pierce-stone', later contracted to parsley piert. It has no relation to the true parsley plant.

It is still given as an infusion either alone or together with other similarly-acting herbs for infections of the kidneys and bladder.

Meadowsweet

Filipendula ulmaria

There are few more graceful plants than the meadowsweet, whose stem, up to four feet in height, bears the creamy white feathery flowers so well known to lovers of the countryside. Though it inhabits the temperate regions of Europe and Asia it still seems peculiarly English.

Its delicate almond-like scent made it one of the favourite herbs for strewing over the floors of medieval halls, especially during festivals and weddings. As a result, one of its old English names is bridewort.

Formerly it was used as a medicine. A drink made from the tops boiled in wine was supposed to help relieve attacks of malaria, while a tea made from them was taken to increase the flow of urine.

For eye-drops or an eye-bath to relieve itching or burning and to clear the sight, put a few flowers into a bowl, pour boiling water over them, then leave to cool and strain.

Mugwort

Artemisia vulgaris

Mugwort may be found growing by roadsides, on rubbish-heaps, along hedge-bottoms and by the banks of streams. It may be recognized by the erect, branched, purplish-red stems which grow five to six feet high with divided leaves, dark green on the upper surface and downy underneath. Its erect clusters of flowers appear, each in its small oval head, in July and August.

Before hops were brought to England in the early sixteenth century mugwort was dried and used to flavour ale, and was made into a hot drink before tea became a universal beverage.

Its principal uses have been as a remedy for obstructed menstruation, especially when given in combination with other herbs such as pennyroyal. Bathing inflamed parts of the body with mugwort and following this with a poultice made with chickweed helps to reduce the inflammation.

Stinging Nettle

Urtica dioica

One of the commonest plants along the sides of field paths, under hedgerows and as a garden weed is the common or stinging nettle. Its single stalks rising from the surrounding grass to a height of up to four feet, with its finely-marked saw-toothed leaves and clusters of white or green flowers, are well known. One needs only to touch a leaf to be stung and the sting, though much less severe than that of the bee or wasp, is still uncomfortable and irritating.

It ought to be used much more than it is for as a vegetable it is both appetizing and nourishing. The tops taken off in spring, cleaned and boiled with onions and stock, make a broth with a meaty taste which, because of its very dark, almost black colour, was called in some parts of the north of England, 'nettle coal'. In 1661 Pepys tasted this, pronounced it good and called it 'nettle porridge'. With broccoli, sprouts and rice it makes a savoury pudding and, fermented with sugar and other ingredients, both wine and beer.

The Romans, who are believed to have introduced the plant into Britain, are said to have rubbed it on their limbs and even whipped each other with its stalks against the damp and cold and as an aid to circulation. 'Urtication', or flogging with nettles, was once regarded as a remedy for rheumatism. The substance injected by this process was formic acid, which is also found in the sting of the bee, recommended in some quarters today. It is said that there is nothing new under the sun!

Rheumatism and kidney trouble were the main illnesses treated with infusions of the stinging nettle. Homoeopathic physicians still employ the diluted tincture as a dressing for burns and one, Dr Burnett, who earned a reputation in London by giving it internally for stone in the kidney and for gout, came to be known as Dr Urtica.

Its uses are many. The fibrous stalk of the nettle was used by the Germans during the First World War to make a durable cloth. It was spun and woven in Scotland before flax replaced it as a textile fibre and the poet Campbell declared that it was more durable than any species of linen.

The juice of the nettle will curdle milk and can be used as a substitute for rennet. In Russia a yellow dye was made from the juice extracted from the root. Nettles also made good fodder for cattle because, when dried, the leaves lose their power to sting. The herb even had a reputation as a hair restorer if the juice from the boiled plant was strained and rubbed regularly into the scalp.

Incidentally, if you are stung by a nettle, don't rub in the sting; just pass a sharp knife blade lightly over the affected part and the little acid-bearing hairs will be drawn out.

Nettle Soup

Cook 3 oz washed young nettle tops and leaves over a low heat in a saucepan, adding no liquid until quite tender. Allow to cool.

Chop and sauté an onion in butter or oil; add ¾ lb peeled and diced potatoes. Sauté again, then add 2–2½ pints boiling stock. When the potatoes are cooked add the nettles with a teaspoon of marjoram, basil or sage, and one of lovage, fresh or dried.

Simmer for 15 minutes, then add ½ oz butter and 2 tablespoons cream. Re-heat before serving, but do not boil.

Deadly Nightshade

Atropa belladonna

Year after year warnings of the poisonous nature of this plant appear in print; year after year it claims its young victims.

It is the most attractive and the most accessible of all poisonous herbs. Its black berries, which are often taken for blackcurrants by children with no knowledge of plant and leaf forms, make it the deadliest of all poisonous herbs. Its botanical name *atropa* is derived from the Greek *Atropos*, the eldest of the three fates, who used the dreaded shears to cut short the thin-spun thread of human life, *belladonna* possibly because Italian ladies were once accustomed to use the juice as an eye-drop to enlarge the pupils and make the eyes more beautiful.

Yet this most dreaded of plants is one of the most widely used and efficient of medicines. In 1833 its important alkaloid atropine was isolated, and this is used in eye-drops, to treat attacks of asthma and spells of bed-wetting, especially in children. Belladonna plasters and liniments are well-known applications for neuralgia, gout, rheumatism and sciatica. Homoeopathic physicians use it internally for the sudden onset of chills with high temperature, for influenza when accompanied by inflamed eyes and throat and for the prevention and treatment of scarlet fever.

The Plantains

Plantago major – Common Plantain
Plantago lanceolata – Ribwort

Everybody knows the general form of these troublesome little weeds which take root in the crevices in walls, alongside paths, on waste ground and especially in lawns where their spreading rosettes of basal leaves, tough and numerous, choke the grass around. These are the two most common of the five members of the species found in Great Britain and for them, little boys go out into the fields to collect the spikes on which the seeds are closely packed together, as food for their cagebirds.

The common plantain had a great reputation in the past for healing wounds and ulcers. Romeo alludes to it as an application for mending a broken shin. An ointment was made from it and used as a remedy for piles. An infusion of the seeds strained and made into a syrup with sugar or honey was given to children who had thrush.

The ribwort plantain is now being grown for medicinal purposes on a commercial scale. Its juice, like that of the common variety, is applied to wounds, and tea, syrups and lozenges are given for chills and chest colds.

Self-Heal

Prunella vulgaris

Self-heal is one of the common plants in English pasture lands in July and August when its deep purple heads on square stalks rise among the grass.

Its reputation lies rather in the past than the present. Growing near to cottage and workshop it was always at hand to be crushed and laid on fresh wounds to stop the bleeding. The corolla being the shape of a sickle gave it the common name of sicklewort – an illustration of the Doctrine of Signatures.

There are not many plants which give a double illustration of this doctrine, but since the flower of self-heal bore a resemblance to the shape of the human throat, the infusion was used as a

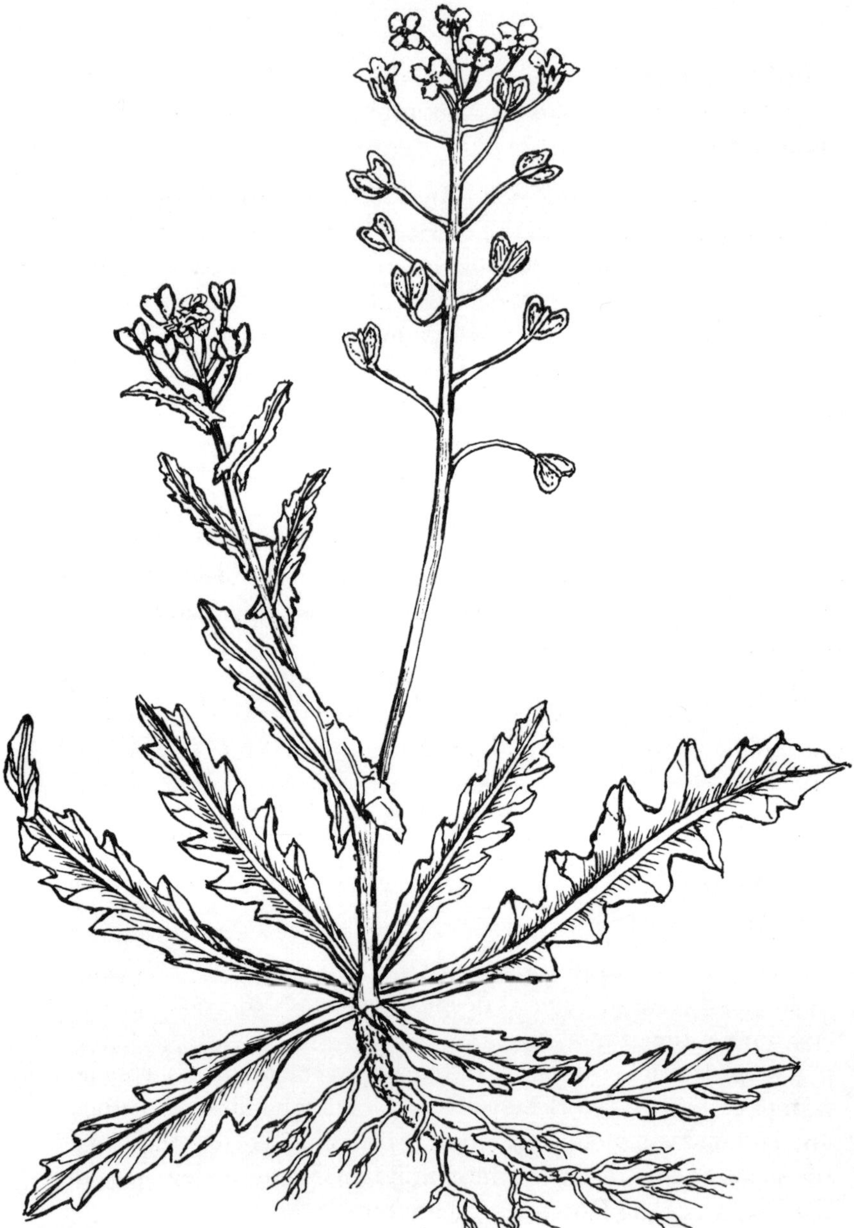

Shepherd's Purse
Capsella bursa pastoris

gargle for inflammation of the mouth, tonsillitis and quinsy. It is still used by herbalists as an astringent, a gargle and a mouthwash.

Shepherd's Purse

Capsella bursa pastoris

Throughout the year, except in the depth of winter, this little plant is to be found in the poorest soil in almost any waste place, sheltering under walls, by roadsides and even in niches between stones wherever its seeds fall.

It bears small white flowers but is most recognizable by the rosette of green leaves at its base and by the diminutive fruit which is heart-shaped and resembles the purse or wallet of the medieval shepherd. It is a native of Europe but has found a home in all parts of the world.

Once known as Poor Man's Pharmacetie, it was widely used both externally and internally for checking bleeding and reducing inflammation. Its chief use today by herbal practitioners is as an infusion to help check bleeding from the kidneys and bladder.

Sweet Flag

Acorus calamus

Like all water-loving plants, sweet flag or sweet sedge, as it is sometimes called, grows best in the neighbourhood of ponds, slow-flowing streams and marshes. It was brought to Europe from Asia in the fifteenth century, has since spread throughout the Continent and was introduced into England by the herbalist John Gerard about 1596. It is also found wild in the United States.

Sweet flag grows from a root or rhizome which can be split for the propagation of new plants, and its long sword-shaped leaves give it a resemblance to the iris. Its flowering shoot sends out a blunt spike of very small greenish-yellow sweet-scented florets. When bruised it has an agreeable and rather orange-like scent.

The important part of the plant is its root which is ground and used as a powder, and from which an infusion is made to stimulate the appetite and aid digestion. From the dried root a drug is extracted which acts in the same way. The oil derived from it is used in improving the flavour of gin and in making perfume.

Herbs of Foreign Origin

extracts of which are used in Great Britain

Aloe *Aloe vera*

A succulent plant native to east and south Africa, cultivated in the West Indies. Aloes is the expressed juice of the leaves, and is purgative and bitter.

Arrowroot *Maranta arundinacea*

A tropical plant cultivated in many places including the West Indies, Bengal, Java, the Philippines and west Africa. A nutritious starch is obtained from the roots and this furnishes an easily-digested convalescent diet. It can also be used to thicken sauces.

Asafoetida *Ferula asafoetida*

An umbelliferous plant growing in Afghanistan and eastern Iran. The resin extracted by incision of the root has a bitter acrid taste. A tonic for flatulence and a stimulant to the respiratory and intestinal tracts.

Benzoin *Styrax tonkinensis*

A tree growing in Thailand and Indonesia. The resin is collected from incisions in the trunk and makes a stimulant, antiseptic tincture. This tincture, known as Friar's Balsam, is used as an inhalant for inflammation of the respiratory tract.

Bindweed Jalap *Convolvulus jalapa*

A bindweed with small turnip-shaped tubers originating in the Mexican Andes, now also cultivated in India. The tubers yield a resin which is strongly purgative and is used in cases of constipation, especially if liver trouble is involved. Jalap was given frequently to soldiers during the First World War in such cases.

Bitter Ash *Picraena excelsa*

A tree resembling the ash which is a native of the West Indies. Its wood, sold in chips or raspings, is used as an insecticide, a bitter tonic and as an enema for expelling worms. The familiar name of the commodity is quassia.

Cacao *Theobroma cacao*

A low tree native to tropical America and Africa from the seeds of which cocoa and chocolate are derived.

Californian Buckthorn *Rhamnus purshiana*

A tree growing on the Pacific coast of North America, now cultivated for its bark from which a liquid extract is made, known as cascara sagrada. This, acting as a laxative, is also sold in tablet form.

Camphor *Cinnamomum camphora*

A tree native to China, Japan and eastern Asia. Chips from the tree, steamed, provide common camphor, an antiseptic and counter-irritant used in liniments.

Cardamoms *Elletaria cardamomum*

A large perennial herb of southern India. The dried ripe seeds, aromatic and spicy, are an ingredient of curry powder.

Castor Oil *Ricinus communis*

A plant grown in tropical and subtropical conditions, also grown in Great Britain, but rarely flowering here. The seeds, pressed, provide the well-known oil used for cases of temporary constipation.

Cinchona *Cinchona succirubra*

First discovered in South America, this evergreen tree yields a bark which when powdered was found to cure fevers. Its name is believed to have been derived from the Countess of Cinchon, wife of a Viceroy of Peru who was cured by it. It was commonly known as Peruvian or Jesuit's Bark and from it the alkaloid quinine was first extracted in 1820.

Coca *Erythroxylon coca*

A small tree, native to Bolivia and Peru, from whose leaves cocaine, a stimulant, local anaesthetic and drug, is obtained.

Coffee *Coffea arabica*

A tropical tree originally from south-west Ethiopia but now cultivated in many subtropical parts of the world. The seeds, roasted and ground, constitute coffee and contain the drug caffeine, a stimulant for the nerves and vascular system.

Eucalyptus *Eucalyptus globulus*

A tree native to Australia, now also grown in North and South Africa, India and southern Europe. From the leaves is distilled an oil which is a powerful antiseptic used as an inhalant for colds and bronchial troubles.

Ginseng *Panax schinseng*

A perennial herb cultivated in eastern Asia, Korea and Japan. It had a great reputation in China for preserving youth and vitality. The roots and those of the American ginseng (*Panax quinquefolia*) provide a tincture. The herb is now being widely advertised in the West for peculiar virtues but its pharmaceutical and therapeutic value is so far doubtful.

Golden Seal *Hydrastis canadensis*

A small perennial herb growing in Canada and the eastern United States. The root provides a tonic and laxative.

Henna *Lawsonia inermis*

A dwarf shrub growing in Egypt and western Asia. The flowers distilled in water once provided a perfume. The leaves and young twigs reduced to a fine powder are used as a dye for the hair and, in some parts of the world, for the nails and skin.

Indian Hemp *Apocynum cannabinum, Cannabis sativa*

An annual grown in India which furnishes the drug cannabis in several forms – *bhang, ganja, hashish* and *cannabis resin.*

Indigo *Indigofera tinctoria*

An annual plant grown in India and other subtropical countries which furnishes a blue dyestuff, the use of which is now almost entirely superseded by chemical dyes.

Ipecacuanha *Cephaëlis ipecacuanha*

A shrubby plant, native to South America. An extract of the root consisting of various alkaloids is used to stimulate appetite and improve digestion. It is also incorporated into wine.

Liquorice *Glycyrrhiza glabra*

A leguminous plant growing mainly in south-eastern Europe and south-western Asia. The roots, cleaned and dried, used to be sold in confectioners' shops for chewing. The black extract of liquorice used to be called 'Spanish Juice'. It is incorporated into many kinds of confectionery and cough medicines. Liquorice powder was once a favourite mild purgative and cough remedy for young children.

Nux Vomica *Strychnos nux-vomica*

A moderate-sized tree flourishing in India and Malaysia. Strychnine is obtained from the dried ripe seeds, and a tincture from the seeds is mixed with water to make a bitter tonic and appetizer. Great care is needed in administering it, for taken in quantity it is a poison.

Pokeroot *Phytolacca decandra*

A perennial herb growing in North America and in Mediterranean countries. A tincture is made from the dried roots and is used as a slow purgative. Homoeopaths use the drug in a strong solution as a powerful remedy for sore and inflamed throats.

White Poppy *Papaver somniferum*

An annual plant growing widely between Asia Minor and China. Opium is extracted from the poppy heads. Derivatives are the tincture laudanum, morphine (which is given to kill pain, induce sleep and so on), and the dangerous addictive drug heroin.

Sarsaparilla *Smilax*

A perennial climbing plant native to Central America. Extracts of the root once had a wide reputation as a blood purifier. It is perhaps best known because of the drink of the same name, of which it forms the main ingredient.

Sassafras *Sassafras officinale*

A tree growing in north-east America between Canada and Mexico. From the bark and the root is extracted an aromatic oil used for flavouring beverages, confectionery and so on.

Senna *Cassia acutifolia*

A small shrub growing in north-east Africa and Arabia. Its dried leaves and pods are used as a decoction for constipation. Various confections are made from it for domestic use.

Slippery Elm *Ulmus fulva*

A small tree growing in the United States and Canada. The inner bark, finely powdered, is used as a soothing nutritive food, especially for those with stomach and bowel complaints. Ground more coarsely it is used in poultices for abscesses, inflammations and old wounds.

Snakeroot, Canadian Snakeroot *Asarum canadense*

A small plant whose root is dried and given for colds and chronic chest complaints, and to induce perspiration. Virginia snakeroot (*aristolochia serpentaria*) is a perennial herb whose dried roots are given in small doses as a stimulant and tonic.

Tea *Camellia sinensis*

The small evergreen shrub whose leaves, dried and cleaned, provide this universally-popular stimulating beverage. First grown in Ceylon and China, it is now to be found in various subtropical countries.

Tobacco *Nicotiana tabacum*

An annual plant, produced in Virginia and later in many subtropical countries. Its leaves are dried and cured to provide smoking mixture and snuff. It is grown to a very limited extent in the British Isles. A local irritant, it is not used for medicinal purposes, and it is becoming increasingly clear that smoking tobacco can severely damage the health.

Tragacanth *Astragalus gummifera*

A small thorny shrub, a native of Asia Minor and Iran. It provides a gummy exudation which is used to give consistency to lozenges containing other substances.

Turmeric *Curcuma longa*

A herbaceous perennial plant found in southern Asia and the West Indies. The dried rhizomes are ground and used in curry powders and as a colouring agent.

Wintergreen *Pyrola rotundifolia*

This small evergreen plant is a member of the heath family which grows in the eastern United States and southern Canada and can be grown in England. Its leaves have an astringent and aromatic odour. From them a volatile oil is produced for external application in cases of muscular stiffness and rheumatism.

Witch-Hazel *Hamamelis virginica* grows as a shrub in the eastern United States and Canada. It will grow in England but is rare. A distilled extract from the young leaves is used for application on piles, bruises and inflammation, bites, burns and scalds. On a lint bandage it is used to dress troublesome varicose veins.

Index